Student Study Guide

New York, New York Columbus, Ohio Chicago, Illinois

McGraw-Hill Glencoe

The McGraw-Hill Companies

Send all inquiries to:
The McGraw-Hill Companies
8787 Orion Place
Columbus, OH 43240-4027

ISBN-13: 978-0-07-877249-8
ISBN-10: 0-07-877249-4

Core-Plus Mathematics
Contemporary Mathematics in Context
Course 1 Student Study Guide

Printed in the United States of America.

15 MAL 15 14

Table of Contents

Math Skills Study Guide

Number Sense Basics
Fractions 1–10
Mixed Numbers 11–12
Decimals 13–16
Scientific Notation 17–18
Absolute Value 19–20
Integers 21–24

Algebra Basics
Ratios, Rates, and Proportions 25–32
Percent 33–36
Order of Operations 37–38
Variables and Expressions 39–40
Inequalities 41–42
The Coordinate Plane 43–44
Functions and Linear Equations 45–46

Geometry Basics
Angles and Bisectors 47–52
Triangles 53–54
Quadrilaterals 55–56
Symmetry 57–58
Translations and Reflections 59–62
Perimeter, Area, and Surface Area 63–74
Measurement Conversion 75
Scale Drawings 76–77

Statistics and Probability
Measures of Center 78–81
Probability 82–85
Organizing Data 86–90

Spiral Review

Unit 1
Lesson 1 91
Lesson 2 92
Lesson 3 93
Lessons 1, 2, and 3 94

Unit 2
Lesson 1 95–96
Lesson 2 97
Lessons 1 and 2 98

Unit 3
Lesson 1 99
Lesson 2 100
Lesson 3 101
Lessons 1, 2, and 3 102–104

Unit 4
Lesson 1 105
Lessons 1 and 2 106

Unit 5
Lesson 1 107–108
Lesson 2 109–110

Unit 6
Lesson 1 111
Lesson 2 112
Lesson 3 113
Lessons 1, 2, and 3 114

Unit 7
Lesson 1 115
Lesson 2 116
Lesson 3 117
Lessons 1, 2, and 3 118

Unit 8
Lesson 1 119–120
Lesson 2 121–122

Standardized Test Practice

Test Practice 123–138

Name ____________________ Date ________ Period ________

Math Skills Study Guide

Simplifying Fractions

Fractions that have the same value are called **equivalent fractions.** The **Greatest Common Factor (GCF)** of two numbers is the largest number that can be divided into both numbers. A fraction is in **simplest form** when the GCF of the numerator and denominator is 1.

EXAMPLE 1 **Write $\frac{36}{54}$ in simplest form.**

First, find the GCF of the numerator and denominator.
factors of 36: 1, 2, 3, 4, 6, 9, 12, 18, 36
factors of 54: 1, 2, 3, 6, 9, 18, 27, 54
The GCF of 36 and 54 is 18.

Then, divide the numerator and the denominator by the GCF.

$\frac{36}{54} = \frac{36 \div 18}{54 \div 18} = \frac{2}{3}$ So, $\frac{36}{54}$ written in simplest form is $\frac{2}{3}$.

EXAMPLE 2 **Write $\frac{8}{12}$ in simplest form.**

$8 = 2 \cdot 2 \cdot 2$

$12 = 2 \cdot 2 \cdot 3$

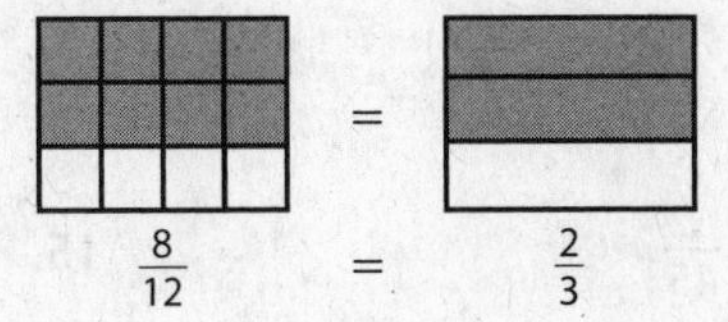

$\frac{8}{12} = \frac{2}{3}$

GCF: $2 \cdot 2 = 4$

$\frac{8}{12} = \frac{8 \div 4}{12 \div 4} = \frac{2}{3}$

So, $\frac{8}{12}$ written in simplest form is $\frac{2}{3}$.

EXERCISES

Write each fraction in simplest form.

1. $\frac{42}{72}$ **2.** $\frac{40}{54}$ **3.** $\frac{21}{35}$

4. $\frac{25}{100}$ **5.** $\frac{99}{132}$ **6.** $\frac{17}{85}$

Name ________________________ Date ______ Period ______

Math Skills Study Guide

Simplifying Fractions

Write each fraction in simplest form.

1. $\frac{49}{70}$ **2.** $\frac{5}{30}$ **3.** $\frac{6}{14}$

4. $\frac{14}{28}$ **5.** $\frac{72}{72}$ **6.** $\frac{18}{21}$

7. $\frac{45}{75}$ **8.** $\frac{50}{200}$ **9.** $\frac{32}{50}$

10. $\frac{56}{64}$ **11.** $\frac{14}{35}$ **12.** $\frac{39}{45}$

13. $\frac{48}{66}$ **14.** $\frac{42}{45}$ **15.** $\frac{78}{130}$

Write two fractions that are equivalent to each fraction.

16. $\frac{3}{4}$ **17.** $\frac{7}{9}$ **18.** $\frac{7}{11}$

19. $\frac{14}{17}$ **20.** $\frac{21}{23}$ **21.** $\frac{11}{17}$

Name ______________________ Date ________ Period ________

Math Skills Study Guide

Adding and Subtracting Fractions with Like Denominators

Fractions with the same denominator are called **like fractions.**

- To add like fractions, add the numerators. Use the same denominator in the sum.
- To subtract like fractions, subtract the numerators. Use the same denominator in the difference.

EXAMPLE 1 **Find the sum of $\frac{3}{5}$ and $\frac{3}{5}$.**

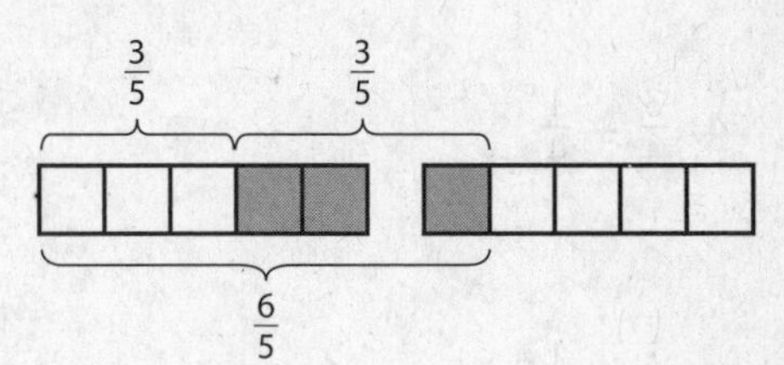

Estimate $\frac{1}{2} + \frac{1}{2} = 1$

$\frac{3}{5} + \frac{3}{5} = \frac{3+3}{5}$ Add the numerators.

$= \frac{6}{5}$ Simplify.

Compared to the estimate, the answer is reasonable.

EXAMPLE 2 **Find the difference of $\frac{3}{4}$ and $\frac{1}{4}$.**

Estimate $1 - 0 = 1$

$\frac{3}{4} - \frac{1}{4} = \frac{3-1}{4}$ Subtract the numerators.

$= \frac{2}{4}$ or $\frac{1}{2}$ Simplify.

Compared to the estimate, the answer is reasonable.

EXERCISES

Add or subtract. Write in simplest form.

1. $\frac{1}{9} + \frac{4}{9}$ **2.** $\frac{9}{11} - \frac{7}{11}$ **3.** $\frac{9}{10} + \frac{5}{10}$ **4.** $\frac{11}{12} - \frac{9}{12}$

5. $\frac{4}{7} + \frac{5}{7}$ **6.** $\frac{4}{9} - \frac{1}{9}$ **7.** $\frac{7}{8} + \frac{5}{8}$ **8.** $\frac{6}{7} - \frac{4}{7}$

9. $\frac{3}{4} + \frac{3}{4}$ **10.** $\frac{4}{5} - \frac{1}{5}$ **11.** $\frac{5}{6} + \frac{1}{6}$ **12.** $\frac{7}{10} - \frac{1}{10}$

Name ______________________________ Date ________ Period ________

Math Skills Study Guide

Adding and Subtracting Fractions with Like Denominators

Add or subtract. Write in simplest form.

1. $\frac{2}{9} + \frac{4}{9}$

2. $\frac{2}{5} + \frac{4}{5}$

3. $\frac{2}{3} - \frac{1}{3}$

4. $\frac{3}{4} + \frac{1}{4}$

5. $\frac{7}{8} - \frac{3}{8}$

6. $\frac{9}{12} + \frac{3}{12}$

7. $\frac{5}{6} - \frac{1}{6}$

8. $\frac{1}{6} + \frac{5}{6}$

9. $\frac{11}{12} - \frac{7}{12}$

10. $\frac{7}{8} + \frac{3}{8}$

11. $\frac{9}{10} - \frac{4}{10}$

12. $\frac{3}{8} + \frac{1}{8}$

13. $\frac{10}{11} - \frac{2}{11}$

14. $\frac{7}{9} + \frac{2}{9}$

15. $\frac{5}{6} + \frac{4}{6}$

16. $\frac{3}{10} - \frac{1}{10}$

17. $\frac{3}{10} + \frac{3}{10}$

18. $\frac{5}{6} + \frac{3}{6}$

19. $\frac{5}{8} - \frac{3}{8}$

20. $\frac{5}{7} - \frac{2}{7}$

21. $\frac{6}{7} + \frac{5}{7}$

22. How much is $\frac{2}{9}$ pound plus $\frac{1}{9}$ pound?

23. How much longer is $\frac{3}{8}$ foot than $\frac{1}{8}$ foot?

24. How much more than $\frac{1}{4}$ cup is $\frac{3}{4}$ cup?

25. What is the sum of $\frac{2}{11}$, $\frac{7}{11}$, and $\frac{1}{11}$?

Name ______________________ Date ________ Period ________

Math Skills Study Guide

Adding and Subtracting Fractions with Unlike Denominators

The **Least Common Denominator (LCD)** is the smallest number that is divisible by both denominators. To find the sum or difference of two fractions with unlike denominators, rename the fractions using the LCD. Then add or subtract and simplify.

EXAMPLE 1 **Find $\frac{1}{3} + \frac{5}{6}$.**

The LCD of $\frac{1}{3}$ and $\frac{5}{6}$ is 6.

Write the problem.

$$\frac{1}{3} + \frac{5}{6}$$

→

Rename $\frac{1}{3}$ as $\frac{2}{6}$.

$$\frac{1}{3} \times \frac{2}{2} = \frac{2}{6}$$

$$\frac{5}{6}$$

→

Add the fractions.

$$\frac{2}{6} + \frac{5}{6} = \frac{7}{6} \text{ or } 1\frac{1}{6}$$

EXAMPLE 2 **Find $\frac{2}{3} - \frac{1}{4}$.**

The LCD of $\frac{2}{3}$ and $\frac{1}{4}$ is 12.

Write the problem.

$$\frac{2}{3} - \frac{1}{4}$$

→

Rename $\frac{2}{3}$ as $\frac{8}{12}$ and $\frac{1}{4}$ as $\frac{3}{12}$.

$$\frac{2}{3} \times \frac{4}{4} = \frac{8}{12}$$

$$\frac{1}{4} \times \frac{3}{3} = -\frac{3}{12}$$

→

Subtract the fractions.

$$\frac{8}{12} - \frac{3}{12} = \frac{5}{12}$$

EXAMPLE 3 **Evaluate $x - y$ if $x = \frac{1}{2}$ and $y = \frac{2}{5}$.**

$x - y = \frac{1}{2} - \frac{2}{5}$ — Replace x with $\frac{1}{2}$ and y with $\frac{2}{5}$.

$= \frac{1}{2} \times \frac{5}{5} - \frac{2}{5} \times \frac{2}{2}$ — Rename $\frac{1}{2}$ and $\frac{2}{5}$ using the LCD, 10.

$= \frac{5}{10} - \frac{4}{10}$ — Simplify.

$= \frac{1}{10}$ — Subtract the numerators.

EXERCISES

Add or subtract. Write in simplest form.

1. $\frac{1}{6} + \frac{1}{2}$

2. $\frac{2}{3} - \frac{1}{2}$

3. $\frac{1}{4} + \frac{7}{8}$

4. $\frac{9}{10} - \frac{3}{5}$

5. $\frac{2}{7} + \frac{1}{2}$

6. $\frac{5}{6} - \frac{1}{12}$

7. $\frac{7}{10} + \frac{1}{2}$

8. $\frac{4}{9} - \frac{1}{3}$

9. Evaluate $x + y$ if $x = \frac{1}{12}$ and $y = \frac{1}{6}$.

10. Evaluate $a + b$ if $a = \frac{1}{2}$ and $b = \frac{3}{4}$.

Name ______________________ Date ________ Period ________

Math Skills Study Guide

Adding and Subtracting Fractions with Unlike Denominators

Add or subtract. Write in simplest form.

1. $\frac{2}{3} + \frac{5}{6}$

2. $\frac{5}{6} + \frac{3}{4}$

3. $\frac{2}{3} - \frac{1}{6}$

4. $\frac{1}{2} + \frac{7}{8}$

5. $\frac{4}{7} - \frac{1}{2}$

6. $\frac{1}{6} - \frac{1}{12}$

7. $\frac{5}{8} - \frac{1}{4}$

8. $\frac{1}{3} + \frac{5}{7}$

9. $\frac{1}{5} + \frac{5}{6}$

10. $\frac{3}{4} + \frac{11}{12}$

11. $\frac{1}{2} - \frac{2}{5}$

12. $\frac{11}{12} - \frac{3}{4}$

13. $\frac{3}{4} - \frac{1}{12}$

14. $\frac{4}{5} + \frac{1}{2}$

15. $\frac{3}{5} + \frac{2}{3}$

16. $\frac{2}{3} - \frac{1}{4}$

17. $\frac{11}{12} - \frac{1}{6}$

18. $\frac{3}{5} + \frac{9}{10}$

19. How much more is $\frac{3}{8}$ gallon than $\frac{1}{4}$ gallon?

20. How much more is $\frac{3}{4}$ ounce than $\frac{1}{3}$ ounce?

21. Evaluate $x - y$ if $x = \frac{7}{10}$ and $y = \frac{3}{5}$.

22. Evaluate $s + t$ if $s = \frac{2}{3}$ and $t = \frac{5}{6}$.

Name ______________________ Date ________ Period ________

Math Skills Study Guide

Multiplying Fractions

Type of Product	What To Do	Example
two fractions	Multiply the numerators. Then multiply the denominators.	$\frac{2}{3} \times \frac{4}{5} = \frac{2 \times 4}{3 \times 5} = \frac{8}{15}$
fraction and a whole number	Rename the whole number as an improper fraction. Multiply the numerators. Then multiply the denominators.	$\frac{3}{11} \times 6 = \frac{3}{11} \times \frac{6}{1} = \frac{18}{11} = 1\frac{7}{11}$

EXAMPLE 1 **Find $\frac{2}{5} \times \frac{3}{4}$.** **Estimate:** $\frac{1}{2} \times 1 = \frac{1}{2}$

$\frac{2}{5} \times \frac{3}{4} = \frac{2 \times 3}{5 \times 4}$ Multiply the numerators. Multiply the denominators.

$= \frac{6}{20}$ or $\frac{3}{10}$ Simplify. Compare to the estimate.

EXAMPLE 2 **Find $\frac{4}{9} \times 8$.** **Estimate:** $\frac{1}{2} \times 8 = 4$

$\frac{4}{9} \times 8 = \frac{4}{9} \times \frac{8}{1}$ Write 8 as $\frac{8}{1}$.

$= \frac{4 \times 8}{9 \times 1}$ Multiply.

$= \frac{32}{9}$ or $3\frac{5}{9}$ Simplify. Compare to the estimate.

EXAMPLE 3 **Find $\frac{2}{5} \times \frac{3}{8}$.** **Estimate:** $\frac{1}{2} \times \frac{1}{2} = \frac{1}{4}$

$\frac{2}{5} \times \frac{3}{8} = \frac{\overset{1}{\cancel{2}} \times 3}{5 \times \underset{4}{\cancel{8}}}$ Divide both the numerator and denominator by the common factor, 2.

$= \frac{3}{20}$ Simplify. Compare to the estimate.

EXERCISES

Multiply. Write in simplest form.

1. $\frac{1}{4} \times \frac{5}{6}$ **2.** $\frac{3}{7} \times \frac{3}{4}$ **3.** $4 \times \frac{1}{5}$ **4.** $\frac{5}{12} \times 2$

5. $\frac{3}{5} \times 10$ **6.** $\frac{2}{3} \times \frac{3}{8}$ **7.** $\frac{1}{7} \times \frac{1}{7}$ **8.** $\frac{2}{9} \times \frac{1}{2}$

Name________________________ Date________ Period________

Math Skills Study Guide

Multiplying Fractions

Multiply. Write in simplest form.

1. $\frac{3}{4} \times \frac{1}{2}$

2. $\frac{1}{3} \times \frac{2}{5}$

3. $\frac{1}{3} \times 6$

4. $\frac{2}{5} \times \frac{3}{7}$

5. $\frac{3}{8} \times 10$

6. $\frac{1}{6} \times \frac{3}{5}$

7. $\frac{2}{9} \times 3$

8. $\frac{9}{10} \times \frac{5}{4}$

9. $\frac{7}{8} \times \frac{2}{9}$

10. $11 \times \frac{3}{4}$

11. $\frac{5}{6} \times \frac{1}{4}$

12. $\frac{4}{9} \times \frac{2}{3}$

13. $\frac{7}{12} \times \frac{6}{11}$

14. $16 \times \frac{5}{12}$

15. $\frac{4}{9} \times \frac{1}{8}$

16. $\frac{1}{5} \times \frac{10}{11}$

17. $\frac{5}{12} \times \frac{3}{8}$

18. $\frac{1}{10} \times \frac{4}{7}$

19. $21 \times \frac{4}{7}$

20. $\frac{5}{9} \times 18$

21. $\frac{5}{6} \times \frac{8}{9}$

For Exercises 22–24, evaluate each expression if $x = 4$, $y = \frac{2}{3}$, and $z = \frac{1}{4}$.

22. $\frac{3}{8}x$

23. xz

24. $3x$

25. xy

26. $9y$

27. $\frac{1}{3}x$

28. yz

29. $8z$

30. xyz

31. If $a = \frac{6}{7}$, what is $\frac{2}{3}a$?

32. Evaluate st if $s = \frac{3}{8}$ and $t = 24$.

Name ______________________ Date ________ Period ________

Math Skills Study Guide

Dividing Fractions

When the product of two numbers is 1, the numbers are called **reciprocals.**

EXAMPLE 1 **Find the reciprocal of 8.**

Since $8 \times \frac{1}{8} = 1$, the reciprocal of 8 is $\frac{1}{8}$.

EXAMPLE 2 **Find the reciprocal of $\frac{5}{9}$.**

Since $\frac{5}{9} \times \frac{9}{5} = 1$, the reciprocal of $\frac{5}{9}$ is $\frac{9}{5}$.

You can use reciprocals to divide fractions. To divide by a fraction, multiply by its reciprocal.

EXAMPLE 3 **Find $\frac{2}{3} \div \frac{4}{5}$.**

$\frac{2}{3} \div \frac{4}{5} = \frac{2}{3} \times \frac{5}{4}$ Multiply by the reciprocal, $\frac{5}{4}$.

$= \frac{\overset{1}{\cancel{2}}}{3} \times \frac{5}{\underset{2}{\cancel{4}}}$ Divide 2 and 4 by the GCF, 2.

$= \frac{5}{6}$ Multiply numerators and denominators.

EXERCISES

Find the reciprocal of each number.

1. 2 **2.** $\frac{1}{6}$ **3.** $\frac{4}{11}$ **4.** $\frac{3}{5}$

Divide. Write in simplest form.

5. $\frac{1}{3} \div \frac{2}{5}$ **6.** $\frac{1}{9} \div \frac{1}{2}$ **7.** $\frac{2}{3} \div \frac{1}{4}$ **8.** $\frac{1}{2} \div \frac{3}{4}$

9. $\frac{4}{5} \div 2$ **10.** $\frac{4}{5} \div \frac{1}{10}$ **11.** $\frac{5}{12} \div \frac{5}{6}$ **12.** $\frac{9}{10} \div 3$

13. $\frac{3}{4} \div \frac{7}{12}$ **14.** $\frac{9}{10} \div 9$ **15.** $\frac{2}{3} \div \frac{5}{8}$ **16.** $4 \div \frac{7}{9}$

Name________________________ Date______ Period______

Math Skills Study Guide

Dividing Fractions

Find the reciprocal of each number.

1. $\frac{1}{2}$ **2.** $\frac{3}{5}$ **3.** 7 **4.** $\frac{8}{11}$

5. 12 **6.** $\frac{9}{10}$ **7.** $\frac{5}{8}$ **8.** $\frac{3}{10}$

Divide. Write in simplest form.

9. $\frac{5}{6} \div \frac{1}{3}$ **10.** $\frac{9}{10} \div \frac{1}{2}$ **11.** $\frac{1}{2} \div \frac{3}{5}$

12. $8 \div \frac{4}{5}$ **13.** $\frac{7}{12} \div \frac{5}{6}$ **14.** $\frac{9}{10} \div \frac{1}{4}$

15. $\frac{3}{8} \div 9$ **16.** $\frac{9}{10} \div \frac{3}{4}$ **17.** $\frac{2}{5} \div \frac{4}{7}$

18. $15 \div \frac{5}{9}$ **19.** $\frac{6}{7} \div \frac{3}{11}$ **20.** $\frac{1}{9} \div \frac{5}{12}$

21. $\frac{5}{6} \div \frac{5}{12}$ **22.** $\frac{10}{11} \div 5$ **23.** $\frac{7}{9} \div \frac{1}{7}$

24. $\frac{6}{7} \div \frac{8}{9}$ **25.** $\frac{3}{5} \div \frac{9}{11}$ **26.** $5 \div \frac{4}{9}$

Find the value of each expression if $x = \frac{1}{4}$, $y = \frac{3}{5}$, and $z = \frac{2}{3}$.

27. $\frac{x}{y}$ **28.** $\frac{z}{2}$ **29.** $\frac{y}{z}$

30. $\frac{z}{x}$ **31.** $\frac{\frac{1}{3}}{x}$ **32.** $\frac{5}{y}$

Name ______________________ Date ________ Period ________

Math Skills Study Guide

Mixed Numbers and Improper Fractions

The number $2\frac{2}{3}$ is a mixed number. A **mixed number** indicates the sum of a whole number and a fraction. The number $\frac{5}{3}$ is an improper fraction. **Improper fractions** are fractions greater than or equal to 1. Mixed numbers can be written as mixed numbers or as improper fractions.

EXAMPLE 1 **Draw a model for $2\frac{1}{3}$. Then write $2\frac{1}{3}$ as an improper fraction.**

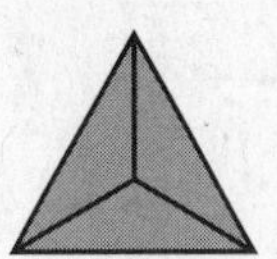
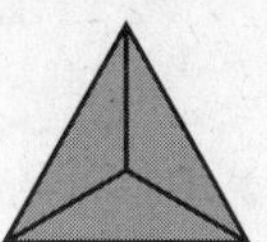
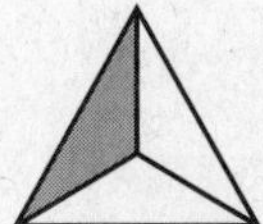

The model shows there are seven $\frac{1}{3}$s.

You can also multiply the denominator and the whole number. Then add the numerator.

$2\frac{1}{3} \rightarrow \frac{(2 \times 3) + 1}{3} = \frac{7}{3}$

So $2\frac{1}{3}$ can be written as $\frac{7}{3}$.

EXAMPLE 2 **Write $\frac{9}{4}$ as a mixed number.**

Divide 9 by 4. Use the remainder as the numerator of the fraction.

$$\begin{array}{r} 2\frac{1}{4} \\ 4\overline{)9} \\ -8 \\ \hline 1 \end{array}$$

So, $\frac{9}{4}$ can be written as $2\frac{1}{4}$.

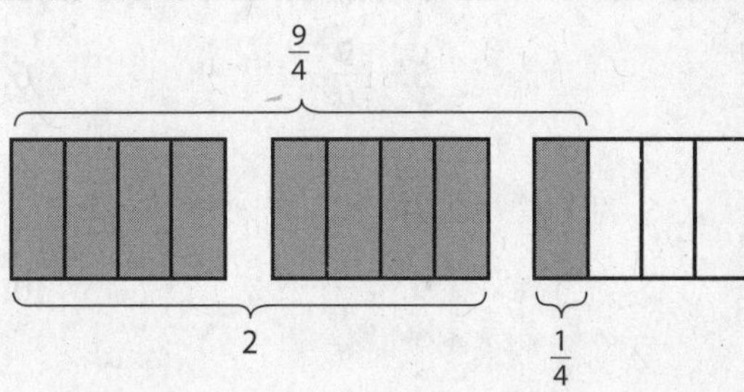

EXERCISES

Write each mixed number as an improper fraction.

1. $3\frac{1}{8}$ **2.** $2\frac{4}{5}$ **3.** $2\frac{1}{2}$ **4.** $1\frac{2}{3}$

5. $2\frac{1}{9}$ **6.** $3\frac{7}{10}$ **7.** $2\frac{3}{8}$ **8.** $1\frac{3}{4}$

Write each improper fraction as a mixed number.

9. $\frac{7}{4}$ **10.** $\frac{5}{3}$ **11.** $\frac{3}{2}$ **12.** $\frac{11}{8}$

13. $\frac{22}{5}$ **14.** $\frac{15}{7}$ **15.** $\frac{25}{4}$ **16.** $\frac{16}{3}$

Name ______________________ Date ________ Period ________

Math Skills Study Guide

Mixed Numbers and Improper Fractions

Draw a model for each mixed number. Then write the mixed number as an improper fraction.

1. $4\frac{1}{3}$

2. $3\frac{3}{8}$

3. $2\frac{2}{5}$

Write each mixed number as an improper fraction.

4. $6\frac{1}{2}$ **5.** $1\frac{5}{6}$ **6.** $1\frac{3}{8}$ **7.** $3\frac{1}{3}$

8. $3\frac{7}{8}$ **9.** $2\frac{1}{4}$ **10.** $2\frac{8}{9}$ **11.** $4\frac{5}{6}$

12. $8\frac{3}{5}$ **13.** $5\frac{4}{7}$ **14.** $10\frac{2}{3}$ **15.** $9\frac{1}{4}$

Write each improper fraction as a mixed number.

16. $\frac{9}{5}$ **17.** $\frac{5}{2}$ **18.** $\frac{15}{4}$ **19.** $\frac{17}{8}$

20. $\frac{19}{6}$ **21.** $\frac{27}{4}$ **22.** $\frac{25}{2}$ **23.** $\frac{31}{7}$

24. $\frac{52}{9}$ **25.** $\frac{41}{3}$ **26.** $\frac{37}{5}$ **27.** $\frac{77}{8}$

Name ______________________ Date ________ Period ________

Math Skills Study Guide

Writing Fractions as Decimals

Any fraction can be written as a decimal using division. Decimals like 0.5 and 0.516 are called **terminating decimals** because the digits end. A decimal like $0.\overline{87} = 0.878787\ldots$ is called a **repeating decimal** because the digits repeat.

EXAMPLE 1 **Write $\frac{3}{8}$ as a decimal.**

Divide.

$$\begin{array}{r} 0.375 \\ 8\overline{)3.000} \\ -2\,4 \\ \hline 60 \\ -56 \\ \hline 40 \\ -40 \\ \hline 0 \end{array}$$

Therefore, $\frac{3}{8} = 0.375$.

EXAMPLE 2 **Write $\frac{7}{11}$ as a decimal.**

Divide.

$$\begin{array}{r} 0.6363 \\ 11\overline{)7.0000} \\ -6\,6 \\ \hline 40 \\ -33 \\ \hline 70 \\ -66 \\ \hline 40 \\ -33 \\ \hline 7 \end{array}$$

The pattern repeats. Therefore, $\frac{7}{11} = 0.\overline{63}$.

EXERCISES

Write each fraction or mixed number as a decimal.

1. $\frac{3}{10}$ **2.** $\frac{3}{4}$ **3.** $\frac{1}{3}$ **4.** $\frac{3}{5}$

5. $\frac{1}{8}$ **6.** $2\frac{1}{4}$ **7.** $1\frac{5}{6}$ **8.** $3\frac{8}{9}$

9. $1\frac{3}{11}$ **10.** $1\frac{5}{8}$ **11.** $3\frac{1}{6}$ **12.** $4\frac{5}{11}$

Name ______________________ Date ________ Period ________

Math Skills Study Guide

Writing Fractions as Decimals

Write each fraction or mixed number as a decimal.

1. $\frac{9}{10}$

2. $\frac{21}{100}$

3. $\frac{3}{4}$

4. $\frac{1}{2}$

5. $\frac{1}{6}$

6. $\frac{5}{6}$

7. $\frac{4}{9}$

8. $3\frac{7}{8}$

9. $9\frac{2}{5}$

10. $\frac{8}{11}$

11. $4\frac{2}{3}$

12. $6\frac{5}{8}$

13. $5\frac{1}{3}$

14. $12\frac{3}{8}$

15. $10\frac{17}{20}$

16. $2\frac{11}{18}$

17. $3\frac{11}{16}$

18. $6\frac{4}{5}$

19. $1\frac{5}{9}$

20. $10\frac{1}{8}$

21. $2\frac{13}{18}$

22. $3\frac{7}{12}$

23. $5\frac{8}{9}$

24. $3\frac{24}{25}$

Name ______________________ Date ________ Period ________

Math Skills Study Guide

Writing Decimals as Fractions

Decimals like 0.58, 0.12, and 0.08 can be written as fractions.
To write a decimal as a fraction, you can follow these steps.

- Identify the place value of the last decimal place.
- Write the decimal as a fraction using the place value as the denominator.
- If necessary, simplify the fraction.

EXAMPLE 1 **Write 0.5 as a fraction in simplest form.**

$0.5 = \frac{5}{10}$ 0.5 means five tenths.

$= \frac{\cancel{5}^{1}}{\cancel{10}_{2}}$ Simplify. Divide the numerator and denominator by the GCF, 5.

$= \frac{1}{2}$ So, in simplest form, 0.5 is $\frac{1}{2}$.

EXAMPLE 2 **Write 0.35 as a fraction in simplest form.**

$0.35 = \frac{35}{100}$ 0.35 means 35 hundredths.

$= \frac{\cancel{35}^{7}}{\cancel{100}_{20}}$ Simplify. Divide the numerator and denominator by the GCF, 5.

$= \frac{7}{20}$ So, in simplest form, 0.35 is $\frac{7}{20}$.

EXAMPLE 3 **Write 4.375 as a mixed number in simplest form.**

$4.375 = 4\frac{375}{1,000}$ 0.375 means 375 thousandths.

$= 4\frac{\cancel{375}^{3}}{\cancel{1,000}_{8}}$ Simplify. Divide by the GCF, 125.

$= 4\frac{3}{8}$

EXERCISES

Write each decimal as a fraction or mixed number in simplest form.

1. 0.9 **2.** 0.8 **3.** 0.27 **4.** 0.75

5. 0.34 **6.** 0.125 **7.** 0.035 **8.** 0.008

9. 1.4 **10.** 3.6 **11.** 6.28 **12.** 2.65

13. 12.05 **14.** 4.004 **15.** 23.205 **16.** 51.724

Name ______________________ Date ________ Period ________

Math Skills Study Guide

Writing Decimals as Fractions

Write each decimal as a fraction or mixed number in simplest form.

1. 0.6	**2.** 10.9	**3.** 0.08
4. 6.25	**5.** 4.125	**6.** 0.075
7. 9.35	**8.** 3.56	**9.** 8.016
10. 21.5	**11.** 0.055	**12.** 7.42
13. 5.006	**14.** 3.875	**15.** 1.29
16. 2.015	**17.** 6.48	**18.** 0.004
19. 4.95	**20.** 8.425	**21.** 9.74
22. 0.47	**23.** 5.019	**24.** 1.062
25. 3.96	**26.** 0.824	**27.** 20.8
28. 6.45	**29.** 4.672	**30.** 0.375

Name ______________________ Date ________ Period ________

Math Skills Study Guide

Scientific Notation

A number is in scientific notation when it is written as the product of a number and a power of ten. The number must be greater than or equal to 1 and less than 10.

- To write a number in standard form, you apply the order of operations. First evaluate the power of ten and then multiply.
- To write a number in scientific notation, move the decimal point to the right of the first nonzero number. Then, find the power of ten by counting the number of places moved.

EXAMPLE 1 **Write 6.1×10^3 in standard form.**

$6.1 \times 10^3 = 6.1 \times 1,000$ — $10^3 = 1,000$

$= 6.1\ 0\ 0$ — Move the decimal point 3 places to the right.

$= 6,100$

EXAMPLE 2 **Write 62,500 in scientific notation.**

$62,500 = 6.250 \times 10,000$ — Move the decimal point 4 places to get a number between 1 and 10.

$= 6.25 \times 10^4$

EXERCISES

Write each number in standard form.

1. 7.25×10^2

2. 2.5×10^3

3. 9.95×10^5

4. 8.80×10^4

5. 3.18×10^6

6. 6.12×10^3

Write each number in scientific notation.

7. 325

8. 9,210

9. 200

10. 5,120

11. 561

12. 1,230

13. 21,300

14. 53,000

15. 8,930

Name ______________________ Date ________ Period ________

Math Skills Study Guide

Scientific Notation

Write each number in standard form.

1. 3.1×10^2

2. 2.3×10^3

3. 9.86×10^2

4. 3.25×10^4

5. 6.10×10^5

6. 7.87×10^4

7. 2.2×10^2

8. 4.27×10^3

9. 1.06×10^7

10. 2.11×10^5

11. 4.82×10^4

12. 5.55×10^{10}

Write each number in scientific notation.

13. 230

14. 300

15. 720

16. 2,790

17. 5,000

18. 8,800

19. 37,000

20. 26,300

21. 52,100

22. 120,000

23. 361,000

24. 989,000

25. 5,000,000

26. 82,100,000

27. 51,000,000

Replace each ● with $<$, $>$, or $=$ to make a true sentence.

28. 3,000 ● 3.0×10^3

29. 520 ● 5.2×10^1

30. 8,800 ● 8.8×10^4

31. 659,000 ● 6.59×10^5

Name ______________________ Date ________ Period ________

Math Skills Study Guide

Integers and Absolute Value

Integers less than zero are **negative integers.** Integers greater than zero are **positive integers.**

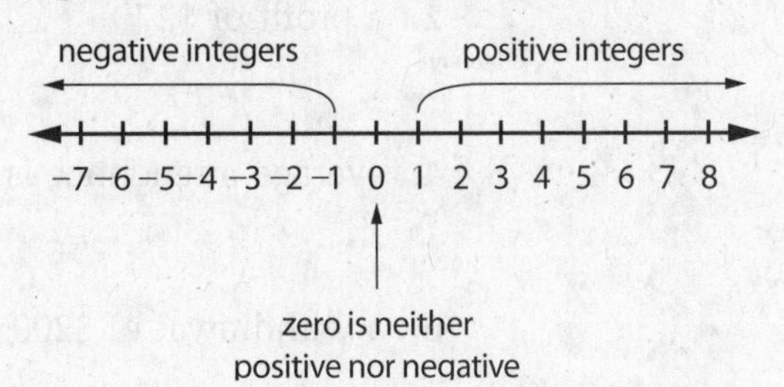

The **absolute value** of an integer is the distance the number is from zero on a number line. Two vertical bars are used to represent absolute value. The symbol for absolute value of 3 is $|3|$.

EXAMPLE 1 **Write an integer that represents 160 feet below sea level.**

Because it represents *below* sea level, the integer is −160.

EXAMPLE 2 **Evaluate $|-2|$.**

On the number line, the graph of −2 is 2 units away from 0. So, $|-2| = 2$.

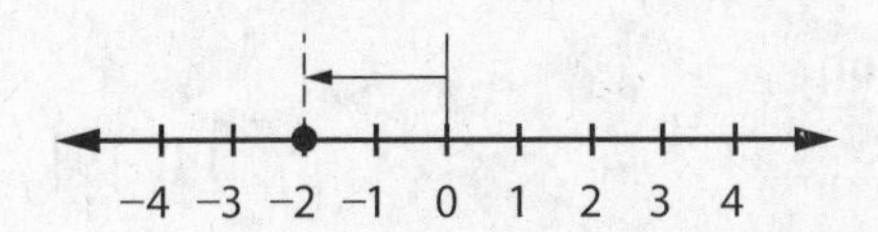

EXERCISES

Write an integer for each situation.

1. 12°C above 0

2. a loss of $24

3. a gain of 20 pounds

4. falling 6 feet

Evaluate each expression.

5. $|12|$

6. $|-150|$

7. $|-8|$

8. $|75|$

9. $|-19|$

10. $|84|$

Name ______________________ Date ________ Period ________

Math Skills Study Guide

Integers and Absolute Value

Write an integer for each situation.

1. 15°C below 0

2. a profit of $27

3. 2010 A.D.

4. average attendance is down 38 people

5. 376 feet above sea level

6. a withdrawal of $200

7. 3 points lost

8. a bonus of $150

9. a deposit of $41

10. 240 B.C.

11. a wage increase of $120

12. 60 feet below sea level

Evaluate each expression.

13. $|-1|$

14. $|9|$

15. $|23|$

16. $|-107|$

17. $|-45|$

18. $|19|$

19. $|0|$

20. $|6| - |-2|$

21. $|-8| + |4|$

22. $|-12| - |12|$

Graph each set of integers on a number line.

23. $\{0, 2, -3\}$

–4 –3 –2 –1 0 1 2 3 4

24. $\{-4, -1, 3\}$

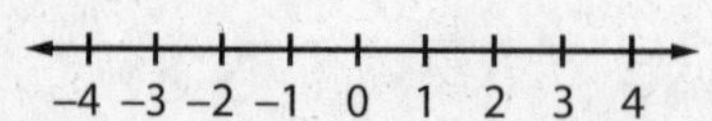

Name ________________________________ Date ________ Period ________

Math Skills Study Guide

Adding Integers

- The sum of two positive integers is always positive.
- The sum of two negative integers is always negative.
- The sum of a positive integer and a negative integer is sometimes positive, sometimes negative, and sometimes zero.

EXAMPLE 1 **Find $-3 + (-2)$.w**

Method 1 Use counters.

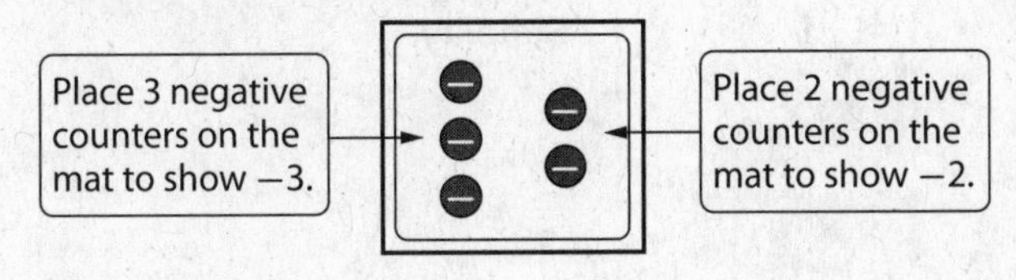

So, $-3 + (-2) = -5$.

Method 2 Use a number line.

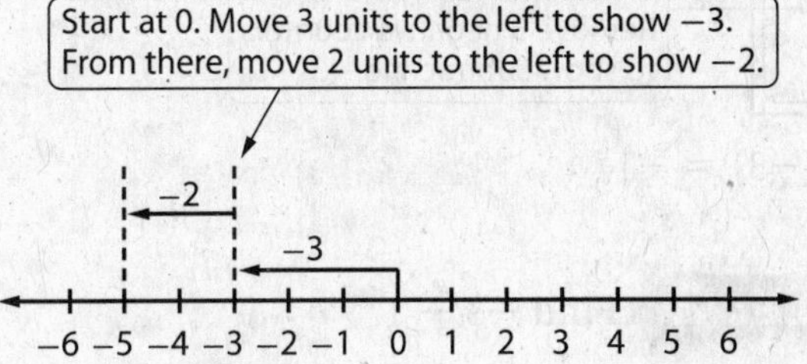

EXAMPLE 2 **Find $4 + (-1)$.**

Method 1 Use counters.

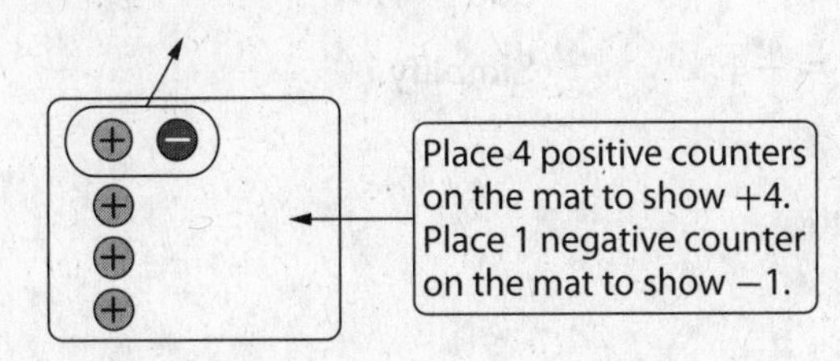

So, $4 + (-1) = 3$.

Method 2 Use a number line.

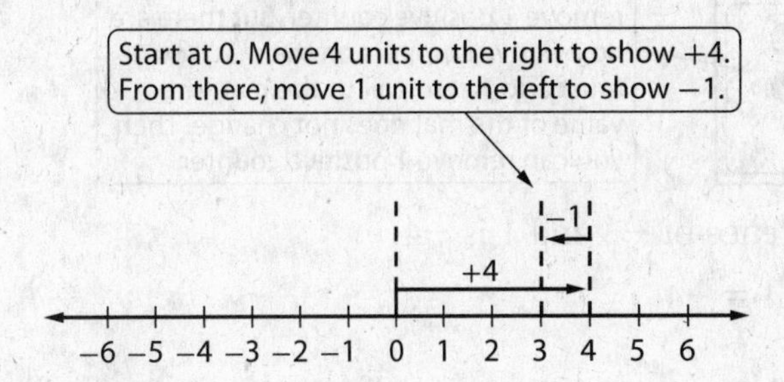

EXERCISES

Add. Use counters or a number line if necessary.

1. $3 + (-6)$

2. $-9 + 8$

3. $-4 + 7$

4. $6 + (-6)$

5. $-8 + (-2)$

6. $2 + (-5)$

7. $6 + (-12)$

8. $-6 + (-5)$

9. $4 + (-3)$

10. $-12 + 5$

11. $-4 + 10$

12. $-3 + (-5)$

Name ______________________ Date ________ Period ________

Math Skills Study Guide

Subtracting Integers

To subtract an integer, change the subtraction symbol to addition, and change the sign of the second integer.

EXAMPLE 1 **Find −4 − (−3).**

Method 1 Use counters.

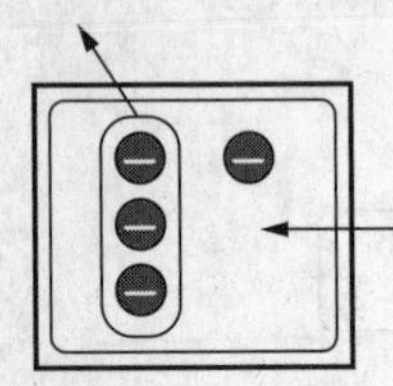

Place 4 negative counters on the mat to show −4. Remove 3 negative counters to show subtracting −3.

So, −4 − (−3) = −1.

Method 2 Use the rule.

$-4 - (-3) = -4 + 3$ To subtract −3, add 3.

$= -1$ Simplify.

EXAMPLE 2 **Find −3 − 1.**

Method 1 Use counters.

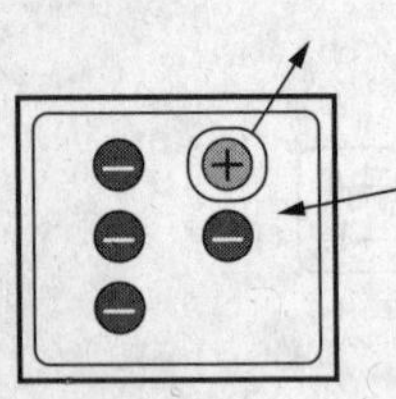

Place 3 negative counters on the mat to show −3. To subtract +1, you must remove 1 positive counter. But there are no positive counters on the mat. You must add 1 zero pair to the mat. The value of the mat does not change. Then you can remove 1 positive counter.

The difference of −3 and 1 is −4.

So, −3 − 1 = −4.

Method 2 Use the rule.

$-3 - 1 = -3 + (-1)$ To subtract 1, add −1.

$= -4$ Simplify.

EXERCISES

Subtract. Use counters if necessary.

1. +8 − 5 **2.** −4 − 2 **3.** 7 − (−5)

4. −3 − (−5) **5.** 6 − (−10) **6.** −8 − (−4)

7. −1 − 4 **8.** 2 − (−2) **9.** −5 − (−1)

10. 7 − 2 **11.** −9 − (−9) **12.** 6 − (−2)

13. −8 − (−14) **14.** −2 − 9 **15.** 5 − 15

Name ______________________ Date ________ Period ________

Math Skills Study Guide

Multiplying Integers

- The product of two integers with different signs is negative.
- The product of two integers with the same sign is positive.

EXAMPLES **Multiply.**

1 **$2 \times (-1)$**

$2 \times (-1) = -2$ The integers have different signs. The product is negative.

2 **-4×3**

$-4 \times 3 = -12$ The integers have different signs. The product is negative.

3 **3×5**

$3 \times 5 = 15$ The integers have the same sign. The product is positive.

4 **$-2 \times (-4)$**

$-2 \times (-4) = 8$ The integers have the same sign. The product is positive.

EXERCISES

Multiply.

1. $3 \times (-3)$

2. $-5 \times (-2)$

3. $-8 \times (-1)$

4. -2×8

5. 4×-3

6. $-3 \times (-2)$

7. $5 \times (-4)$

8. $-10 \times (-4)$

9. -3×6

10. $-3 \times (-10)$

11. $6 \times (-4)$

12. $-7 \times (-7)$

Name ______________________ Date ________ Period ________

Math Skills Study Guide

Dividing Integers

- The quotient of two integers with different signs is negative.
- The quotient of two integers with the same sign is positive.

EXAMPLE 1 **Use counters to find $-6 \div 2$.**

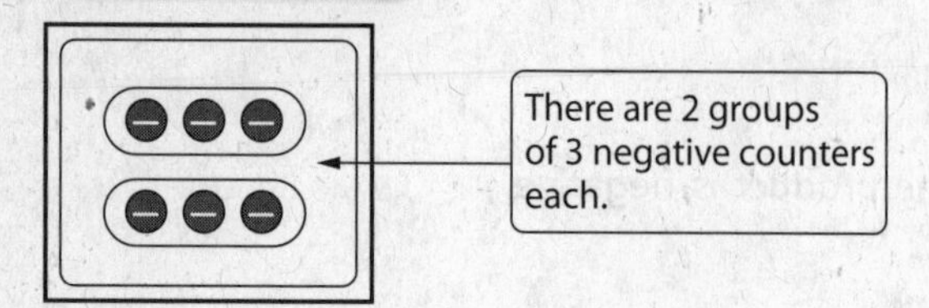

So, $-6 \div 2 = -3$.

EXAMPLES **Divide.**

2 $10 \div (-5)$

Since $-5 \times (-2) = 10$, it follows that $10 \div (-5) = -2$.

3 $-12 \div (-3)$

Since $-3 \times 4 = -12$, it follows that $-12 \div (-3) = 4$.

EXERCISES

Divide.

1. $4 \div (-2)$

2. $-9 \div (-3)$

3. $-8 \div 2$

4. $-21 \div 7$

5. $30 \div (-5)$

6. $-24 \div 4$

7. $-36 \div 6$

8. $-45 \div (-5)$

9. $-81 \div 9$

10. $-3 \div (-3)$

11. $70 \div (-7)$

12. $-64 \div (-8)$

13. **ALGEBRA** Find the value of $a \div b$ if $a = -18$ and $b = 6$.

14. **ALGEBRA** For what value of p is $p \div 5 = -7$ true?

Name ______________________ Date ________ Period ________

Math Skills Study Guide

Ratios

Any ratio can be written as a fraction. To write a ratio comparing measurements, such as units of length or units of time, both quantities must have the same unit of measure. Two ratios that have the same value are **equivalent ratios.**

EXAMPLE 1 **Write the ratio 15 to 9 as a fraction in simplest form.**

$15 \text{ to } 9 = \frac{15}{9}$ Write the ratio as a fraction.

$= \frac{5}{3}$ Simplify.

Written as a fraction in simplest form, the ratio 15 to 9 is $\frac{5}{3}$.

EXAMPLE 2 **Write 40 centimeters to 2 meters as a fraction in simplest form.**

$\frac{40 \text{ centimeters}}{2 \text{ meters}} = \frac{40 \text{ centimeters}}{200 \text{ centimeters}}$ Convert 2 meters to centimeters.

$= \frac{\overset{1}{\cancel{40}} \text{ centimeters}}{\underset{5}{\cancel{200}} \text{ centimeters}}$ Divide by the GCF, 40 centimeters.

$= \frac{1}{5}$ Simplify.

EXERCISES

Write each ratio as a fraction in simplest form.

1. 30 to 12

2. 5:20

3. 49:42

4. 15 to 13

5. 28 feet:35 feet

6. 24 minutes to 18 minutes

7. 75 seconds:2 minutes

8. 12 feet:10 yards

Determine whether the ratios are equivalent. Explain.

9. $\frac{3}{4}$ and $\frac{12}{16}$

10. 12:17 and 10:15

11. $\frac{25}{35}$ and $\frac{10}{14}$

12. 2 lb:36 oz and 3 lb:44 oz

13. 3 ft:12 in. and 2 yd:2 ft

Name ________________ Date ______ Period ______

Math Skills Study Guide

Ratios

Write each ratio as a fraction in simplest form.

1. 14 to 6

2. 18:3

3. 4:22

4. 7:21

5. 18:12

6. 20 to 9

7. 25 to 20

8. 4:10

9. 18:21

10. 84 to 16

11. 33 ounces to 11 ounces

12. 45 minutes:25 minutes

13. 77 cups:49 cups

14. 15 pounds to 39 pounds

15. 40 seconds to 6 minutes

16. 140 centimeters to 3 meters

17. 9 weeks:9 days

18. 1 yard to 11 feet

Determine whether the ratios are equivalent. Explain.

19. $\frac{3}{16}$ and $\frac{9}{48}$

20. $\frac{7}{10}$ and $\frac{8}{11}$

21. 18 in.:3 ft and 12 in.:2 ft

22. 6 mos:2 yr and 8 mos:3 yr

Name ______________________ Date ________ Period ________

Math Skills Study Guide

Ratios and Rates

A **ratio** is a comparison of two numbers by division. A common way to express a ratio is as a fraction in simplest form. Ratios can also be written in other ways. For example, the ratio $\frac{2}{3}$ can be written as 2 to 3, 2 out of 3, or 2:3.

EXAMPLES **Refer to the diagram at the right.**

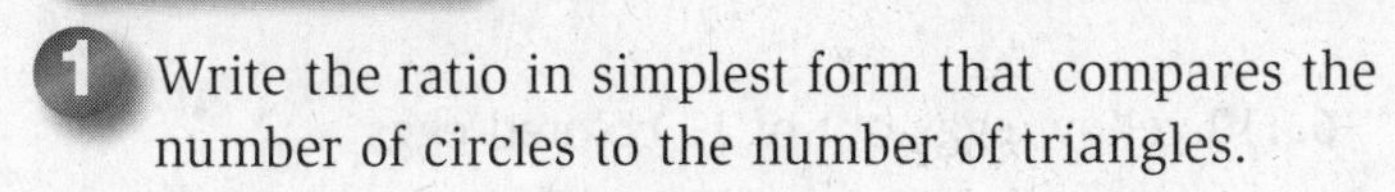

1 Write the ratio in simplest form that compares the number of circles to the number of triangles.

$$\begin{array}{l}\text{circles} \rightarrow \\ \text{triangles} \rightarrow\end{array} \frac{4}{5}$$ The GCF of 4 and 5 is 1.

So, the ratio of circles to triangles is $\frac{4}{5}$, 4 to 5, or 4:5

For every 4 circles, there are 5 triangles.

2 Write the ratio in simplest form that compares the number of circles to the total number of figures.

$$\begin{array}{l}\text{circles} \rightarrow \\ \text{triangles} \rightarrow\end{array} \frac{4}{10} = \frac{2}{5} \quad (\div 2)$$ The GCF of 4 and 10 is 2.

The ratio of circles to the total number of figures is $\frac{2}{5}$, 2 to 5, or 2:5.

For every two circles, there are five total figures.

A **rate** is a ratio of two measurements having different kinds of units. When a rate is simplified so that it has a denominator of 1, it is called a **unit rate.**

EXAMPLE 3 **Write the ratio *20 students to 5 computers* as a unit rate.**

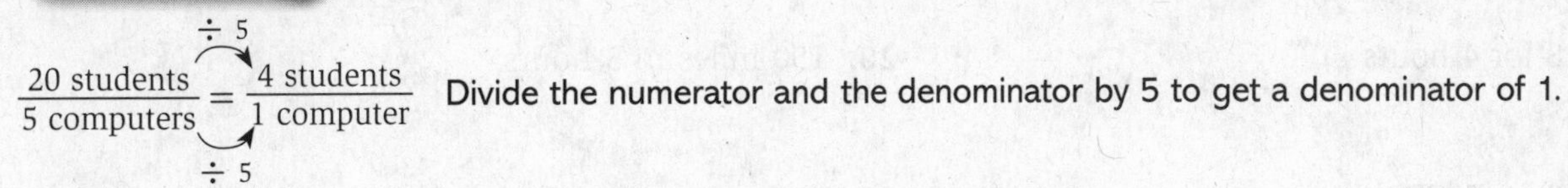

$$\frac{20 \text{ students}}{5 \text{ computers}} = \frac{4 \text{ students}}{1 \text{ computer}} \quad (\div 5)$$ Divide the numerator and the denominator by 5 to get a denominator of 1.

The ratio written as a unit rate is *4 students to 1 computer.*

EXERCISES

Write each ratio as a fraction in simplest form.

1. 2 guppies out of 6 fish
2. 12 puppies to 15 kittens
3. 5 boys out of 10 students

Write each rate as a unit rate.

4. 6 eggs for 3 people
5. $12 for 4 pounds
6. 40 pages in 8 days

Name ______________________ Date ________ Period ________

Math Skills Study Guide

Ratios and Rates

Write each ratio as a fraction in simplest form.

1. 3 sailboats to 6 motorboats

2. 4 tulips to 9 daffodils

3. 5 baseballs to 25 softballs

4. 2 days out of 8 days

5. 6 poodles out of 18 dogs

6. 10 yellow eggs out of 12 colored eggs

7. 12 sheets of paper out of 28

8. 18 hours out of 24 hours

9. 16 elms out of 20 trees

10. 15 trumpets to 9 trombones

11. 5 ducks to 30 geese

12. 14 lions to 10 tigers

13. 6 sodas out of 16 drinks

14. 20 blue jays out of 35 birds

Write each rate as a unit rate.

15. 14 hours in 2 weeks

16. 36 pieces of candy for 6 children

17. 8 teaspoons for 4 cups

18. 8 tomatoes for $2

19. $28 for 4 hours

20. 150 miles in 3 hours

21. $18 for 3 CDs

22. 48 logs on 6 trucks

23. Write the ratio *21 wins to 9 losses* as a fraction in simplest form.

24. Write the ratio *$12 dollars for 3 tickets* as a unit rate.

Name ______________________ Date ________ Period ________

Math Skills Study Guide

Comparing Rates

EXAMPLE 1 **Which is the better deal?**

$35 for 7 balls of yarn; $24 for 4 balls of yarn.

Write each ratio as a fraction. Then find its unit rate.

$$\frac{\$35}{7 \text{ balls of yarn}} \overset{\div 7}{\underset{\div 7}{=}} \frac{\$5}{1 \text{ ball of yarn}} \qquad \frac{\$24}{4 \text{ balls of yarn}} \overset{\div 4}{\underset{\div 4}{=}} \frac{\$6}{1 \text{ ball of yarn}}$$

$35 for 7 balls of yarn is a better deal than $24 for 4 balls of yarn.

EXAMPLE 2 **Which has the greater ratio of boys to students?**

8 boys out of 24 students; 4 boys out of 12 students

Write each ratio as a fraction.

$$\frac{8 \text{ boys}}{24 \text{ students}} \overset{\div 2}{\underset{\div 2}{=}} \frac{4 \text{ boys}}{12 \text{ students}}$$ ← The numerator and the denominator are divided by the same number.

Since the fractions are equivalent, the ratio of boys to the number of students is the same in both cases.

EXERCISES

1. Who saved more per week? $12 saved after 2 weeks; $36 saved after 6 weeks

2. Which is the better deal? $9 for 3 magazines; $20 for 5 magazines

3. Which is the faster driver? 135 miles driven in 3 hours; 225 miles driven in 5 hours

4. Which has more computers per student? 24 computers for 30 students; 48 computers for 70 students

Name ______________________ Date ________ Period ________

Math Skills Study Guide

Comparing Rates

1. Which is the better deal? $18 for 3 bracelets; $30 for 5 bracelets

2. Which has the fewer calories per serving? 120 calories in 2 servings; 360 calories in 6 servings

3. Which is the better pay? 4 hours worked for $12; 7 hours worked for $28

4. Which is the better deal? 15 blank CDs for $5; 45 blank CDs for $15

5. Which scores more points per game? 24 points scored in 4 games; 48 points scored in 10 games

6. Which is the greater fraction of students owning games? 15 out of 20 students own hand-held games; 105 out of 160 students own hand-held games.

7. Which is the faster jog? 30 minutes to jog 3 miles; 50 minutes to jog 5 miles

8. Which is the better deal? $3 for 6 muffins; $9 for 18 muffins

9. Which has the greater gas mileage? 360 miles driven on 12 gallons of fuel; 270 miles driven on 9 gallons of fuel

10. Which is the better deal? 2 pairs of jeans for $50; 4 pairs of jeans for $90

Name ______________________________ Date ________ Period ________

Math Skills Study Guide

Solving Proportions

A **proportion** is an equation stating that two ratios are equivalent. For example, the ratios $\frac{1}{2}$ and $\frac{3}{6}$ are equivalent, so the equation $\frac{1}{2} = \frac{3}{6}$ is a proportion.

$$\frac{1}{2} = \frac{1}{2} \cdot 1$$
$$= \frac{1}{2} \cdot \frac{3}{3}$$
$$= \frac{3}{6}$$

So, $\frac{1}{2} = \frac{3}{6}$.

EXAMPLE 1 Solve $\frac{2}{5} = \frac{4}{n}$.

$\frac{2}{5} \cdot \frac{2}{2} = \frac{4}{10}$

So, $n = 10$.

EXAMPLE 2 Solve $\frac{3}{4} = \frac{b}{12}$.

Multiply both sides by 12,

$12 \cdot \frac{3}{4} = \frac{b}{12} \cdot 12$

$9 = b$

or see that $\frac{3}{4} \cdot \frac{3}{3} = \frac{9}{12}$

So, $b = 9$.

EXERCISES

Determine whether each pair of ratios form a proportion. Explain your reasoning.

1. $\frac{3}{5}, \frac{6}{10}$

2. $\frac{2}{9}, \frac{4}{16}$

Solve each proportion.

3. $\frac{2}{3} = \frac{8}{n}$

4. $\frac{3}{4} = \frac{12}{x}$

5. $\frac{2}{4} = \frac{y}{8}$

6. $\frac{3}{5} = \frac{b}{15}$

7. $\frac{1.2}{9} = \frac{c}{1.5}$

8. $\frac{d}{16} = \frac{3}{8}$

9. $\frac{x}{2.6} = \frac{1.5}{1.3}$

10. $\frac{2}{y} = \frac{6}{9}$

11. $\frac{0.1}{2.6} = \frac{0.5}{z}$

Name ______________________ Date ________ Period ________

Math Skills Study Guide

Solving Proportions

Solve each proportion.

1. $\frac{2}{5} = \frac{8}{x}$

2. $\frac{2}{7} = \frac{4}{y}$

3. $\frac{3}{5} = \frac{b}{30}$

4. $\frac{2}{9} = \frac{c}{36}$

5. $\frac{4}{5} = \frac{d}{25}$

6. $\frac{20}{4} = \frac{10}{f}$

7. $\frac{g}{2} = \frac{28}{14}$

8. $\frac{2}{x} = \frac{10}{25}$

9. $\frac{4}{3} = \frac{h}{18}$

10. $\frac{10}{30} = \frac{2}{r}$

11. $\frac{t}{18} = \frac{3}{6}$

12. $\frac{2}{3} = \frac{6}{m}$

13. $\frac{9}{2} = \frac{s}{6}$

14. $\frac{n}{36} = \frac{2}{6}$

15. $\frac{4}{u} = \frac{12}{21}$

16. $\frac{5}{6} = \frac{m}{12}$

17. $\frac{d}{27} = \frac{4}{9}$

18. $\frac{5}{8} = \frac{15}{q}$

19. $\frac{15}{27} = \frac{5}{k}$

20. $\frac{4}{x} = \frac{20}{30}$

21. $\frac{b}{3} = \frac{24}{9}$

22. $\frac{z}{35} = \frac{4}{7}$

23. $\frac{6}{c} = \frac{24}{28}$

24. $\frac{6}{8} = \frac{x}{24}$

25. $\frac{14}{16} = \frac{b}{8}$

26. $\frac{8}{r} = \frac{24}{27}$

27. $\frac{16}{36} = \frac{t}{9}$

28. $\frac{1.2}{2.4} = \frac{2.4}{n}$

29. $\frac{0.5}{1.8} = \frac{s}{9}$

30. $\frac{1.6}{w} = \frac{8}{16}$

31. What is the solution of $\frac{3}{5} = \frac{2}{k}$? Round to the nearest tenth.

32. Find the solution of $\frac{4.3}{3} = \frac{n}{2.2}$ to the nearest tenth.

Name ______________________ Date ________ Period ________

Math Skills Study Guide

Percent of a Number

You can use a proportion or multiplication to find the percent of a number.

EXAMPLE 1 **SURVEY A survey asked 2,415 people whether they would buy the restored version of The Beatles's *A Hard Day's Night*. 74.95% of the people said they would *not* buy it. How many people would not buy the restored version of this movie?**

$$74.95\% = \frac{74.95}{100}$$

$$= 0.7495$$

$0.7495(2{,}415) = 1{,}810.0425$ Multiply.

So, about 1,810 of the 2,415 people surveyed would not buy the restored version of *A Hard Day's Night*.

EXAMPLE 2 **What number is 15% of 200?**

$15\% \text{ of } 200 = 15\% \times 200$	Write a multiplication expression.
$= 0.15 \times 200$	Write 15% as a decimal.
$= 30$	Multiply.

So, 15% of 200 is 30.

EXERCISES

Find each number.

1. Find 20% of 50.

2. What is 55% of $400?

3. What is 5% of 1,500?

4. Find 190% of 20.

5. What is 24% of $500?

6. 8% of $300 is how much?

7. What is 12.5% of 60?

8. Find 0.2% of 40.

9. Find 3% of $800.

10. What is 0.5% of 180?

11. 0.25% of 42 is what number?

12. What is 0.02% of 280?

Name ______________________ Date ________ Period ________

Math Skills Study Guide

Percent of a Number

Find each number.

1. Find 80% of 80.

2. What is 95% of 600?

3. 35% of 20 is what number?

4. Find 60% of $150.

5. What is 75% of 240?

6. 380% of 30 is what number?

7. Find 40% of 80.

8. What is 30% of $320?

9. 12% of 150 is what number?

10. Find 58% of 200.

11. What is 18% of $450?

12. What is 70% of 1,760?

13. Find 92% of 120.

14. 45% of 156 is what number?

15. What is 12% of 12?

16. Find 60% of 264.

17. 37.5% of 16 is what number?

18. What is 82.5% of 400?

19. What is 0.25% of 900?

20. Find 1.5% of 220.

Name ______________________ Date ________ Period ________

Math Skills Study Guide

Sales Tax and Discount

Sales tax is a percent of the purchase price and is an amount paid in addition to the purchase price. **Discount** is the amount by which the regular price of an item is reduced.

EXAMPLE 1 **SOCCER Find the total price of a $17.75 soccer ball if the sales tax is 6%.**

Method 1

First, find the sales tax.

6% of $17.75 = 0.06 · 17.75
≈ 1.07

The sales tax is $1.07.

Next, add the sales tax to the regular price.

1.07 + 17.75 = 18.82

The total cost of the soccer ball is $18.82.

Method 2

100% + 6% = 106% — Add the percent of tax to 100%.

The total cost is 106% of the regular price.

106% of $17.75 = 1.06 · 17.75
≈ 18.82

EXAMPLE 2 **TENNIS Find the price of a $69.50 tennis racket that is on sale for 20% off.**

First, find the amount of the discount d.

part = percent · base

d = 0.2 · 69.50 — Use the percent equation.

d = 13.90 — The discount is $13.90.

So, the sale price of the tennis racket is $69.50 − $13.90 or $55.60.

EXERCISES

Find the total cost or sale price to the nearest cent.

1. $22.95 shirt; 7% sales tax

2. $39.00 jeans; 25% discount

3. $35 belt; 40% discount

4. $115.48 watch; 6% sales tax

5. $16.99 book; 5% off

6. $349 television; 6.5% sales tax

Name ____________________ Date ________ Period ________

Math Skills Study Guide

Sales Tax and Discount

Find the total cost or sale price to the nearest cent.

1. $49.95 CD player; 5% discount

2. $69 shoes; 6% sales tax

3. $2.99 socks; 5.5% sales tax

4. $119 coat; 40% discount

5. $299 DVD player; 7% sales tax

6. $49 tie; 15% discount

7. $59 power tool; 5% sales tax

8. $17.99 CD; 10% discount

9. $79 cell phone; 20% discount

10. $65 concert ticket; 7.5% sales tax

11. $459 television; 30% discount

12. $19,995 car; 6.5% sales tax

Find the percent of discount to the nearest percent.

13. boots: regular price, $89
sale price, $62.50

14. video game: regular price, $14.99
sale price, $12.64

15. drum set: regular price, $1,240
sale price, $1,099

16. gloves: regular price, $24
sale price, $16.40

17. sweater: regular price, $48
sale price, $34

18. sunglasses: regular price, $80
sale price, $62.95

19. dinner for two: regular price, $75
sale price, $70

20. bicycle: regular price, $189
sale price, $147.85

Name ______________________ Date ________ Period ________

Math Skills Study Guide

Evaluating Expressions Using the Order of Operations

To evaluate an algebraic expression, you replace each variable with its numerical value, then use the **order of operations** simplify.

1. Do all operations within grouping symbols first.
2. Evaluate all powers before other operations.
3. Multiply and divide in order from left to right.
4. Add and subtract in order from left to right.

EXAMPLE 1 **Evaluate $6x - 7$ if $x = 8$.**

$6x - 7 = 6(8) - 7$ Replace x with 8.

$= 48 - 7$ Use the order of operations.

$= 41$ Subtract 7 from 48.

EXAMPLE 2 **Evaluate $5m - 3n$ if $m = 6$ and $n = 5$.**

$5m - 3n = 5(6) - 3(5)$ Replace m with 6 and n with 5.

$= 30 - 15$ Use the order of operations.

$= 15$ Subtract 15 from 30.

EXAMPLE 3 **Evaluate $\frac{ab}{3}$ if $a = 7$ and $b = 6$.**

$\frac{ab}{3} = \frac{(7)(6)}{3}$ Replace a with 7 and b with 6.

$= \frac{42}{3}$ The fraction bar is like a grouping symbol.

$= 14$ Divide.

EXERCISES

Evaluate each expression if $a = 4$, $b = 2$, and $c = 7$.

1. $3ac$

2. $5b^3$

3. abc

4. $5 + 6c$

5. $\frac{ab}{8}$

6. $2a - 3b$

7. $\frac{b^4}{4}$

8. $c - a$

9. $20 - bc$

10. $2bc$

11. $ac - 3b$

12. $6a^2$

Name ______________________ Date ________ Period ________

Math Skills Study Guide

Order of Operations

Evaluate each expression.

1. $9 - 3 + 4$

2. $8 + 6 - 5$

3. $12 \div 4 + 5$

4. $25 \times 2 - 7$

5. $36 \div 9(2)$

6. $6 + 3(7 - 2)$

7. $3 \times 6.2 + 5^2$

8. $(1 + 11)^2 \div 3$

9. $12 - (2 + 8)$

10. $15 - 24 \div 4 \cdot 2$

11. $(4 + 2) \cdot (7 + 4)$

12. $(3 \cdot 18) \div (2 \cdot 9)$

13. $24 \div 6 + 4^2$

14. $3 \times 8 - (9 - 7)^3$

15. $9 + (9 - 8 + 3)^4$

16. $3 \times 2^2 + 24 \div 8$

17. $(15 \div 3)^2 + 9 \div 3$

18. $(52 \div 4) + 5^3$

19. 26×10^3

20. 7.2×10^2

21. $5 \times 4^2 - 3 \times 2$

22. $24 \div 6 \div 2$

23. $13 - (6 - 5)^5$

24. $(8 - 3 \times 2) \times 6$

25. $(11 \cdot 4 - 10) \div 2$

26. $10 \div 2 \times (4 - 3)$

27. 1.82×10^5

28. $35 \div 7 \times 2 - 4$

29. $2^5 + 7(9 - 1)$

30. $12 + 16 \div (3 + 1)$

Name ______________________ Date ________ Period ________

Math Skills Study Guide

Algebra: Variables and Expressions

- A **variable** is a quantity that changes, usually represented by a letter.
- Multiplication in algebra is usually shown as $4n$ for 4 times n.
- **Algebraic expressions** are combinations of variables, numbers, and/or operations.

EXAMPLE 1 **Evaluate $35 + x$ if $x = 6$.**

$35 + x = 35 + 6$ Replace x with 6.

$= 41$ Add 35 and 6.

EXAMPLE 2 **Evaluate $y + x$ if $x = 21$ and $y = 35$.**

$y + x = 35 + 21$ Replace x with 21 and y with 35.

$= 56$ Add 35 and 21.

EXAMPLE 3 **Evaluate $4n + 3$ if $n = 2$.**

$4n + 3 = 4 \times 2 + 3$ Replace n with 2.

$8 + 3$ Find the product of 4 and 2.

11 Add 8 and 3.

EXAMPLE 4 **Evaluate $4n - 2$ if $n = 5$.**

$4n - 2 = 4 \times 5 - 2$ Replace n with 5.

$20 - 2$ Find the product of 4 and 5.

18 Subtract 2 from 20.

EXERCISES

Evaluate each expression if $y = 4$.

1. $3 + y$ **2.** $y + 8$ **3.** $4y$

4. $9y$ **5.** $15y$ **6.** $300y$

7. y^2 **8.** $y^2 + 18$ **9.** $y^2 + 3(7)$

Evaluate each expression if $m = 3$ and $k = 10$.

10. $16 + m$ **11.** $4k$ **12.** mk

13. $m + k$ **14.** $7m + k$ **15.** $6k + m$

16. $3k - 4m$ **17.** $2mk$ **18.** $5k - 6m$

19. $\frac{20m}{k}$ **20.** $m^3 + 2k^2$ **21.** $\frac{k^2}{2 + m}$

Name ________________ Date ________ Period ________

Math Skills Study Guide

Algebra: Variables and Expressions

Complete the table.

Algebraic Expressions	Variables	Numbers	Operations
1. $5d + 2c$			
2. $5w - 4y + 2s$			
3. $xy \div 4 + 3m - 6$			

Evaluate each expression if $a = 3$ and $b = 4$.

4. $10 + b$

5. $2a + 8$

6. $4b - 5a$

7. ab

8. $(7a)(9b)$

9. $8a - 9$

10. $b(22)$

11. $a^2 + 1$

12. $\frac{18}{2a}$

13. a^2b^2

14. $\frac{ab}{3}$

15. $15a - 4b$

16. $ab + 7(11)$

17. $\frac{36}{6a}$

18. $7a + 8b(2)$

Evaluate each expression if $x = 7$, $y = 15$, and $z = 8$.

19. $x + y + z$

20. $x + 2z$

21. $xz + 3y$

22. $4x - 3z$

23. $\frac{z^2}{4}$

24. $6z - 5z$

25. $\frac{9y}{2x + 1}$

26. $15y + x^2$

27. $y^2 + 4(6)$

28. $y^2 - 2x^2$

29. $x^2 + 30 - 18$

30. $13y - \frac{zx}{4}$

31. $xz - 2y + 8$

32. $z^2 + 5y - 20$

33. $3y(40x) - 1,000$

Name ______________________ Date ________ Period ________

Math Skills Study Guide

Inequalities

An **inequality** is a mathematical sentence that contains the symbols $<$, $>$, $\leq$, or $\geq$.

Words	Symbols
m is greater than 7.	$m > 7$
r is less than -4.	$r < -4$
t is greater than or equal to 6.	$t \geq 6$
y is less than or equal to 1.	$y \leq 1$

EXAMPLE 1 **Solve $v + 3 < 5$. Then graph the solution.**

$v + 3 < 5$	Write the inequality.
$-3 \quad -3$	Subtract 3 from each side.
$v < 2$	Simplify.

Check Try 1, a number less than 2.

$v + 3 < 5$	Write the inequality.
$1 + 3 \stackrel{?}{<} 5$	Replace v with 1.
$4 < 5$ ✓	The solution checks.

The solution is all numbers less than 2.

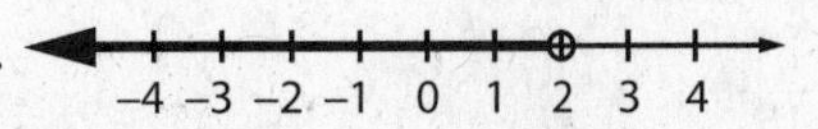

EXAMPLE 2 **Solve $3y + 2 \geq 26$. Then graph the solution.**

$3y + 2 \geq 26$	Write the inequality.
$-2 \quad -2$	Subtract 2 from each side.
$3y \geq 24$	Simplify.
$\frac{3y}{3} \geq \frac{24}{3}$	Divide each side by 3.
$y \geq 8$	Simplify.

Check Use 8 and a number greater than 8.

$3(8) + 2 \stackrel{?}{=} 26$ $\quad$ $3(10) + 2 \stackrel{?}{\geq} 26$

$26 = 26$ ✓ $\quad$ $32 > 26$ ✓

The solution is all numbers greater than or equal to 8.

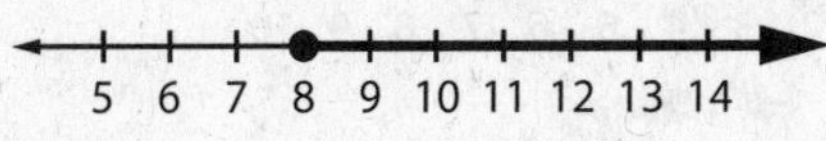

EXERCISES

Graph each inequality on the number line.

1. $c < 5$

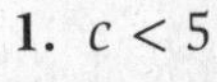

2. $y > -5$

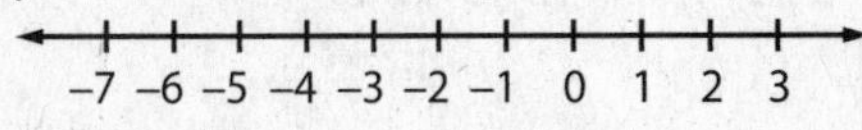

3. $x \geq 10$

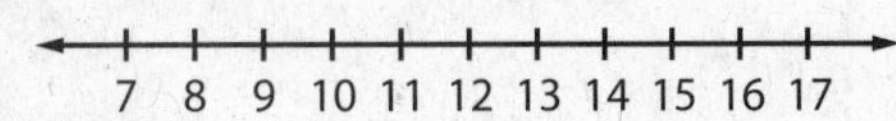

4. $n \leq -1$

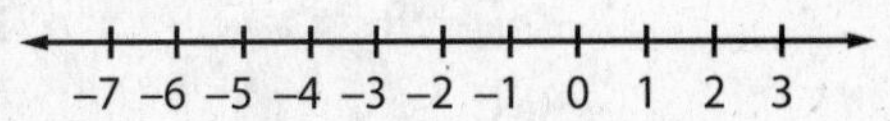

5. $y < 5$

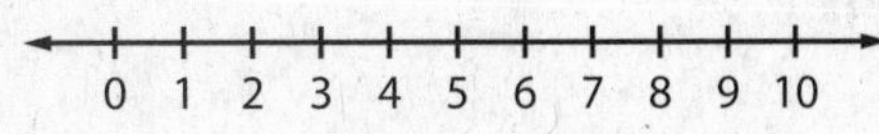

6. $a > 6$

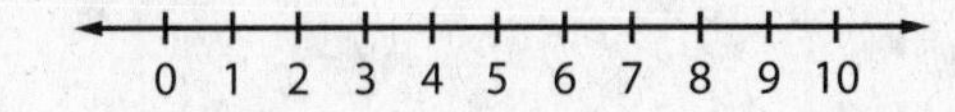

7. $q \leq 8$

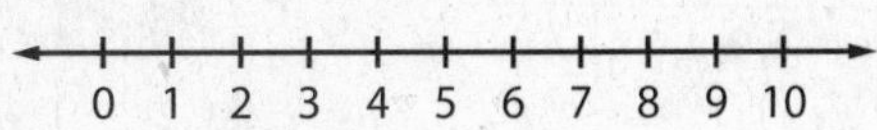

8. $w > 5$

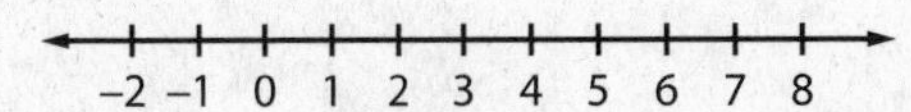

9. $r \leq 7$

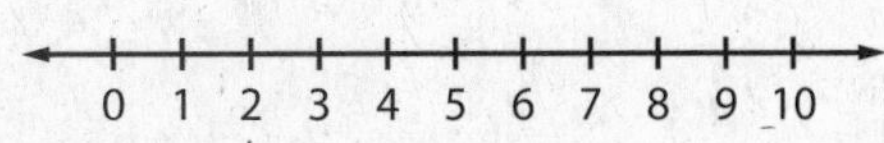

10. $x \geq 1$

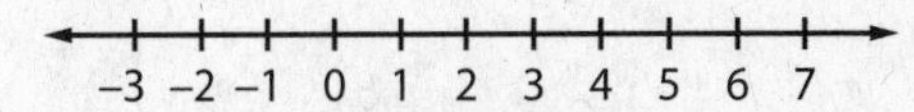

Name ______________________ Date ________ Period ________

Math Skills Study Guide

Inequalities

Graph each inequality on a number line.

1. $x > 2$

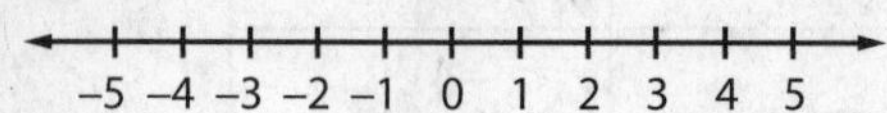

2. $y > -3$

−5 −4 −3 −2 −1 0 1 2 3 4 5

3. $b \geq 1$

−5 −4 −3 −2 −1 0 1 2 3 4 5

4. $c \geq -5$

−5 −4 −3 −2 −1 0 1 2 3 4 5

5. $z < 3$

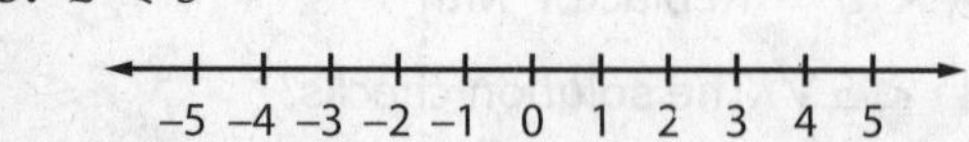

6. $q < -2$

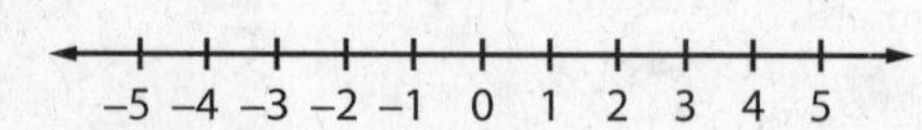

7. $a \leq 6$

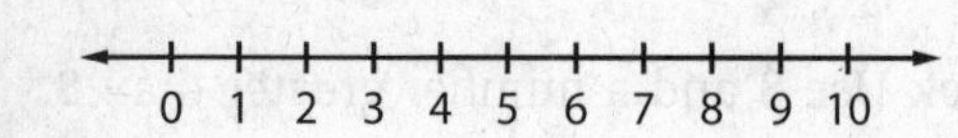

8. $r \leq 0$

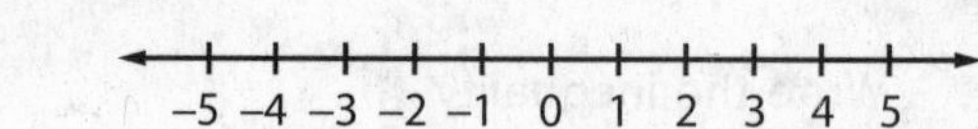

9. $a < 6$

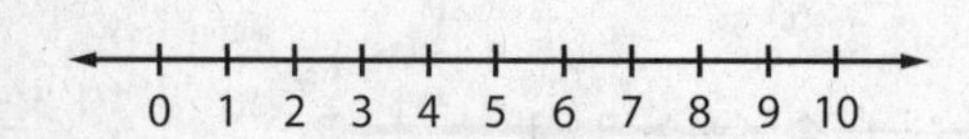

10. $c < 4$

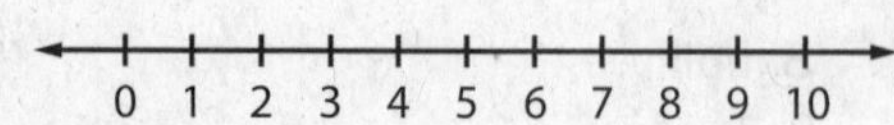

11. $d \geq 9$

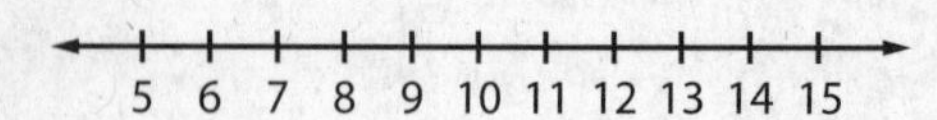

12. $g > 13$

10 11 12 13 14 15 16 17 18 19 20

13. $t \geq -10$

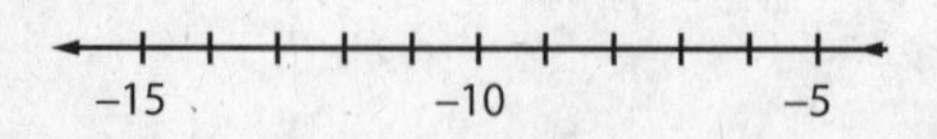

14. $a < -4$

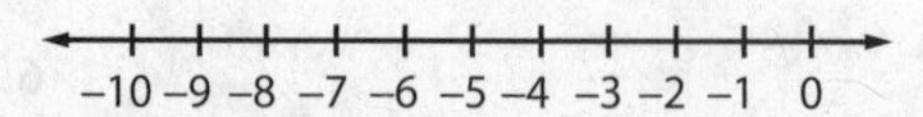

Name ______________________ Date ________ Period ________

Math Skills Study Guide

The Coordinate Plane

The *x*-axis (horizontal) and *y*-axis (vertical) separate the coordinate plane into four regions called **quadrants.**

EXAMPLE 1 **Identify the ordered pair that names point *A*.**

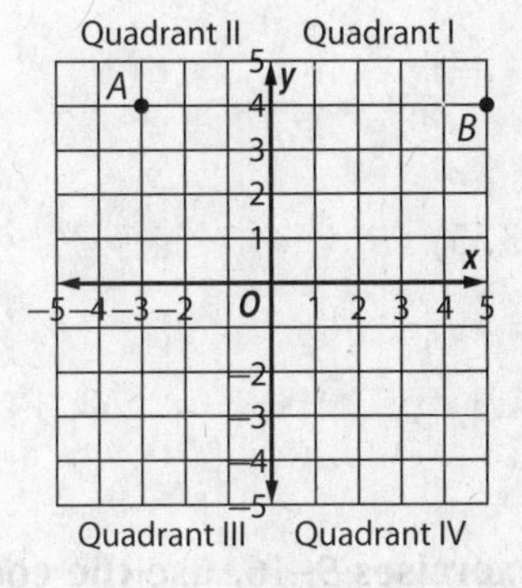

Step 1 From point *A*, trace down to the *x*-axis to find the *x*-coordinate of point *A*, which is –3.

Step 2 From point *A*, trace over to the *y*-axis to find the *y*-coordinate, which is 4.

Point *A* is named by (–3, 4).

EXAMPLE  **Graph point *B* at (5, 4).**

Locate 5 on the *x*-axis.
Locate 4 on the *y*-axis.

Draw a dot that lines up with both.

EXERCISES

Use the coordinate plane at the right. Write the ordered pair that names each point.

1. *C* **2.** *D*

3. *E* **4.** *F*

5. *G* **6.** *H*

7. *I* **8.** *J*

Graph and label each point using the coordinate plane at the right.

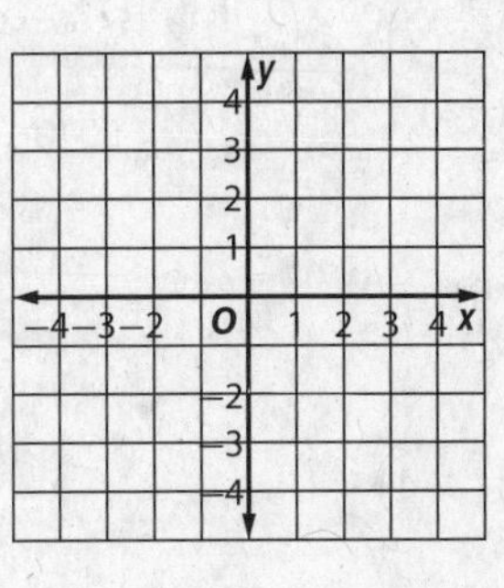

9. $A(-5, 5)$ **10.** $M(2, 4)$

11. $G(0, -5)$ **12.** $D(3, 0)$

13. $N(-4, -3)$ **14.** $I(2, -3)$

Name ______________________________ Date ________ Period ________

Math Skills Study Guide

The Coordinate Plane

For Exercises 1–8, use the coordinate plane at the right. Identify the point for each ordered pair.

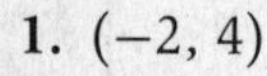

1. (−2, 4)

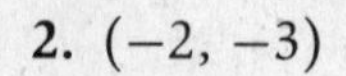

2. (−2, −3)

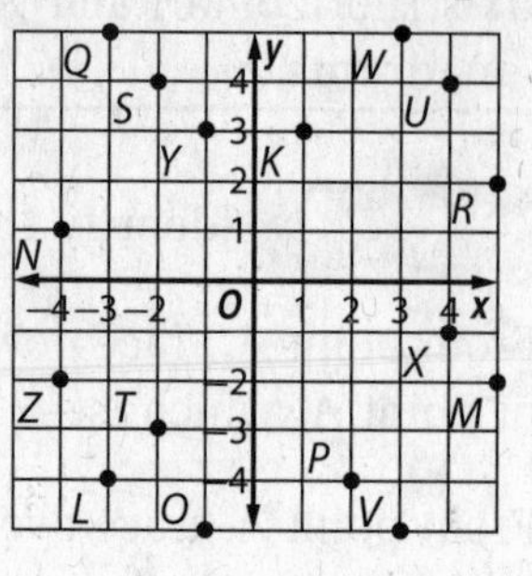

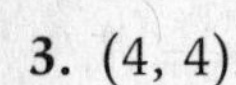

3. (4, 4)

4. (3, −5)

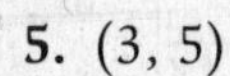

5. (3, 5)

6. (4, −1)

7. (−1, 3)

8. (−4, −2)

For Exercises 9–16, use the coordinate plane above. Write the ordered pair that names each point. Then identify the quadrant where each point is located.

9. *K*

10. *L*

11. *M*

12. *N*

13. *O*

14. *P*

15. *Q*

16. *R*

Graph and label each point on the coordinate plane at the right.

17. *A*(−5, 2)

18. *I*(2, 1)

19. *J*(1, −3)

20. *B*(−5, −1)

21. *C*(3, 3)

22. *K*(−1, 2)

23. *L*(0, −1)

24. *D*(2, −5)

25. *E*(3, −2)

26. *M*(−4, −5)

27. *N*(1, 5)

28. *F*(−2, 5)

29. *G*(−1, −4)

30. *O*(5, −5)

Name ______________________ Date ________ Period ________

Math Skills Study Guide

Linear Functions

A function rule gives an algebraic expression for calculating the value of the output from the input. The output for a given input is usually written as an ordered pair (x, y), which can be graphed. If the graphed (input, output) values for a function form a straight line, then the function is linear, and all points on the line represent (input, output) values that satisfy the function rule.

EXAMPLE 1 **Graph $y = 3x - 2$.**

Select any four values for the input x. We chose 2, 1, 0, and −1. Substitute these values for x to find the output y.

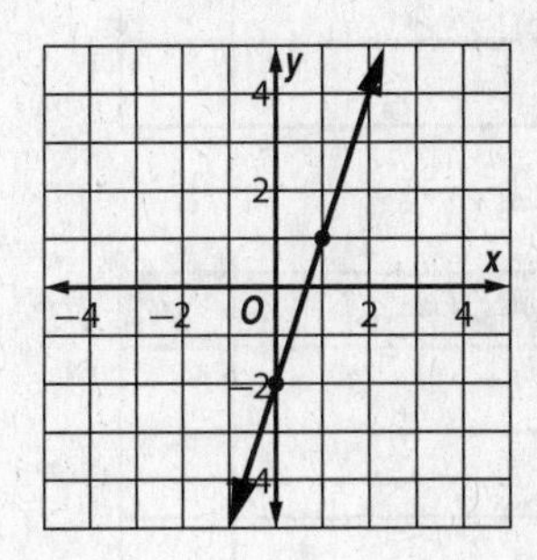

x	$3x - 2$	y	(x, y)
2	$3(2) - 2$	4	(2, 4)
1	$3(1) - 2$	1	(1, 1)
0	$3(0) - 2$	−2	(0, −2)
−1	$3(-1) - 2$	−5	(−1, −5)

Four solutions are (2, 4), (1, 1), (0, −2), and (−1, −5).
The graph is shown at the right.

EXERCISES

Complete a table of sample values and sketch a graph for each function.

1. $y = x + 2$

x	$x + 2$	y

2. $y = -x$

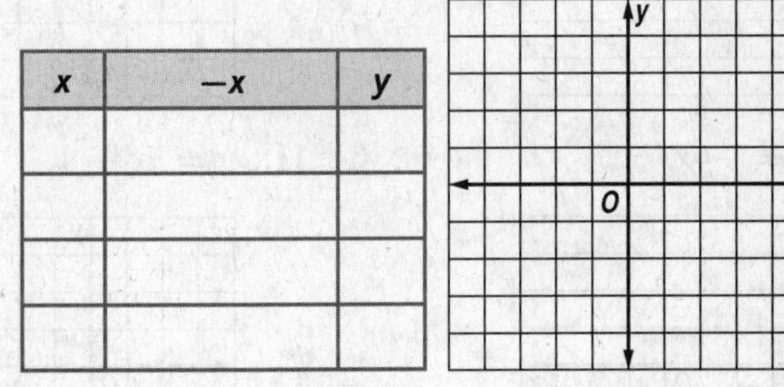

x	$-x$	y

3. $y = 4x$

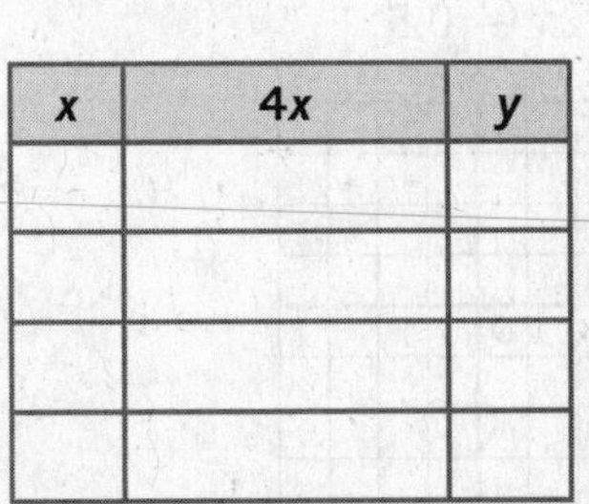

x	$4x$	y

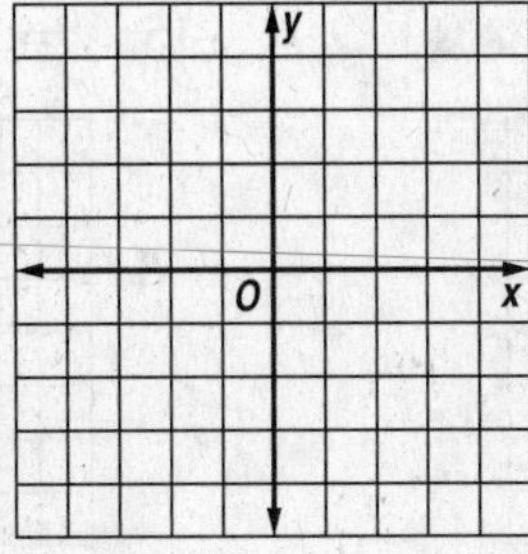

4. $y = 2x + 4$

x	$2x + 4$	y

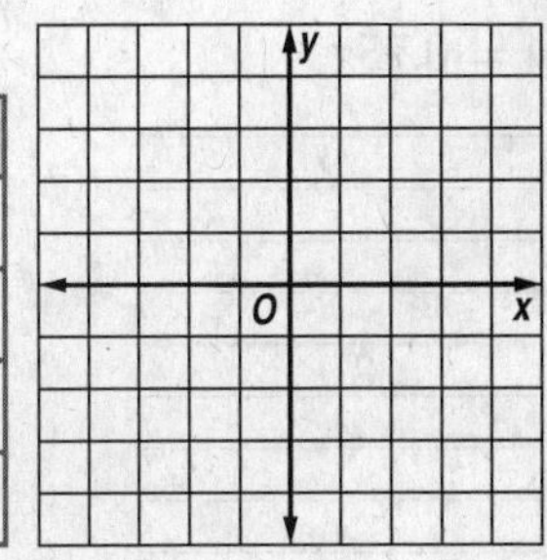

Name ______________________ Date ________ Period ________

Math Skills Study Guide

Linear Functions

Copy and complete each function table.

1. $y = x - 1$

x	$x - 1$	y
1		
2		
3		
4		

2. $y = x + 7$

x	$x + 7$	y
−5		
−3		
−1		
1		

3. $y = 3x$

x	$3x$	y
1		
2		
3		
4		

4. $y = -4x$

x	$-4x$	y
−1		
0		
1		
2		

5. $y = 3x + 1$

x	$3x + 1$	y
−1		
0		
1		
2		

6. $y = -2x + 3$

x	$-2x + 3$	y
−1		
0		
1		
2		

Graph each function.

7. $y = x - 1$

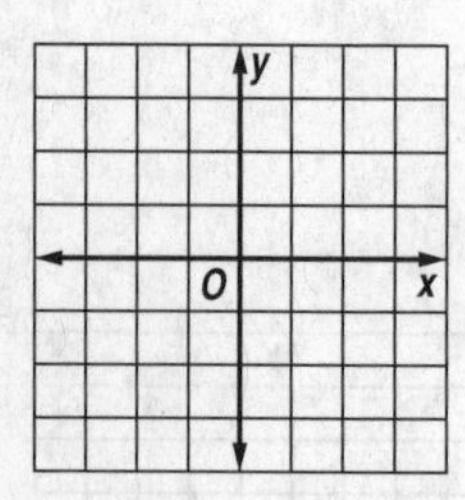

8. $y = x + 7$

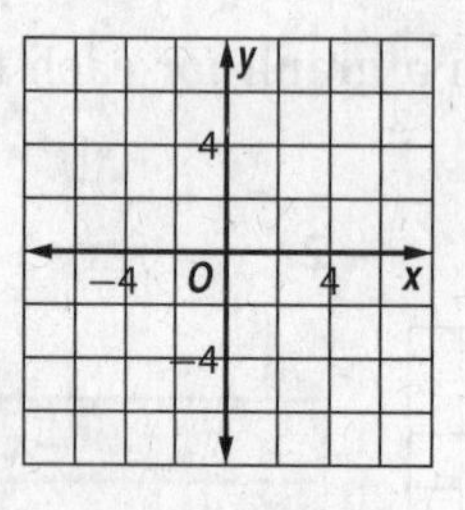

9. $y = 3x$

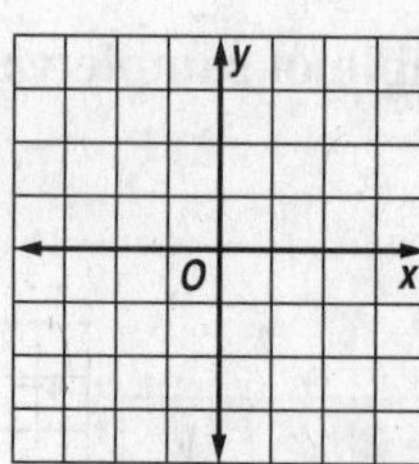

10. $y = -4x$

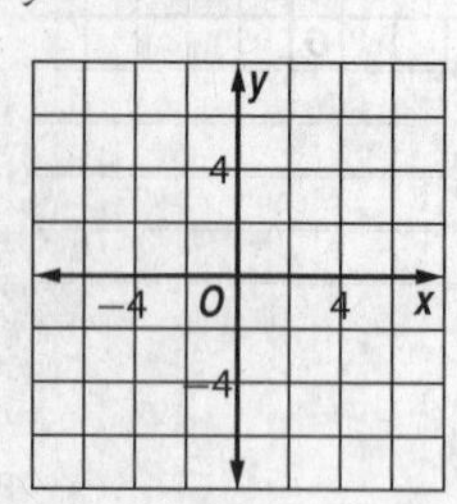

11. $y = 3x + 1$

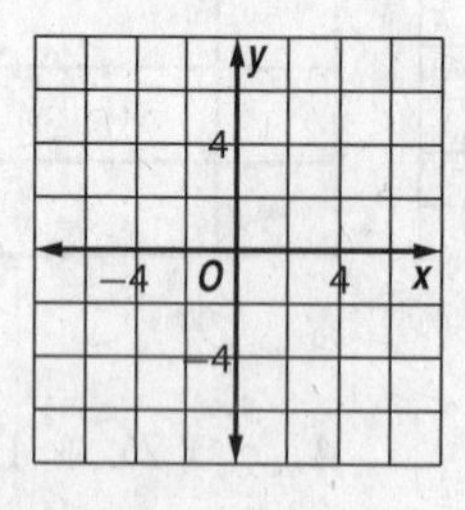

12. $y = -2x + 3$

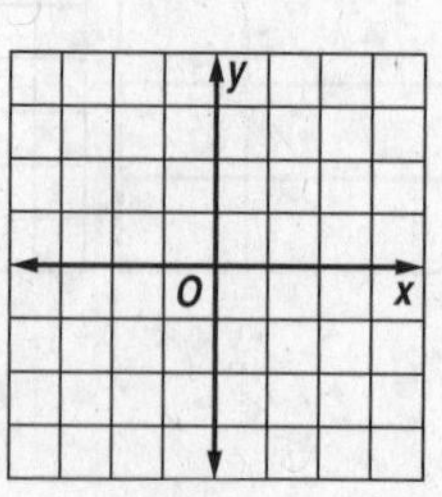

13. $y = 0.75x$

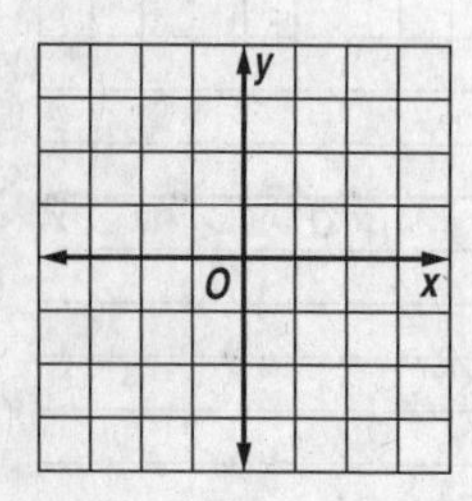

14. $y = 0.5x + 1$

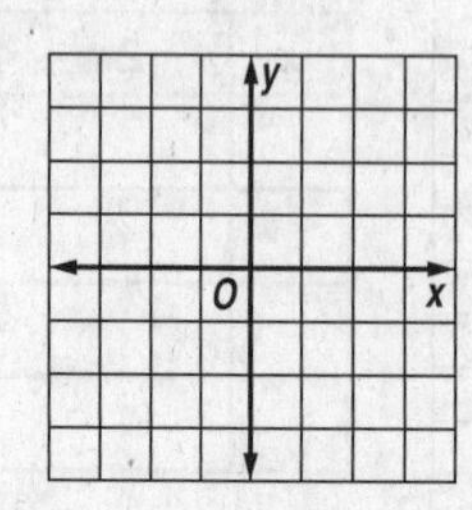

15. $y = 2x - 0.5$

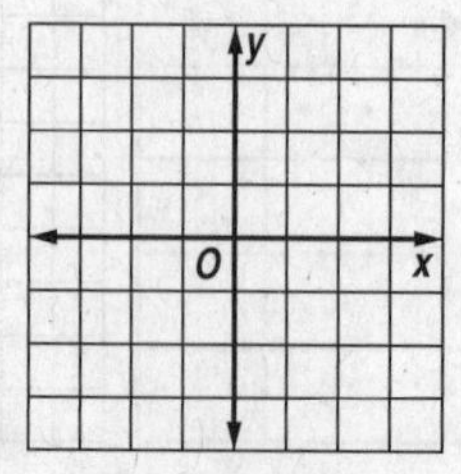

Name ______________________ Date ________ Period ________

Math Skills Study Guide

Angles

Angles have two **sides** that share a common endpoint called the **vertex.**
Angles are measured in **degrees.** One degree is equal to $\frac{1}{360}$th of a circle.
Angles can be classified according to their measure.

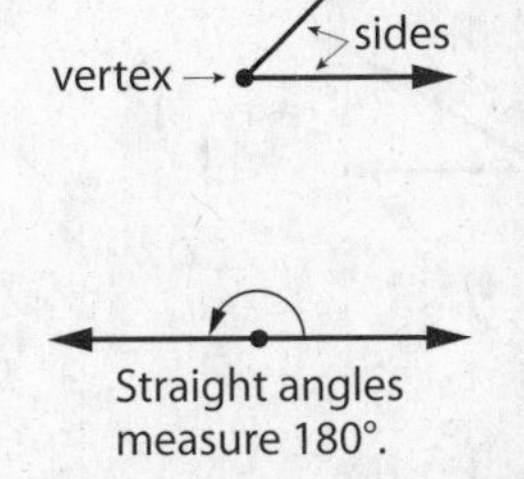

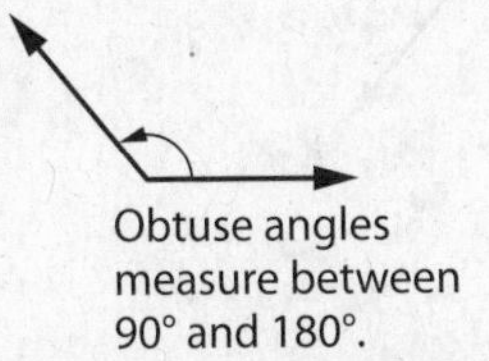

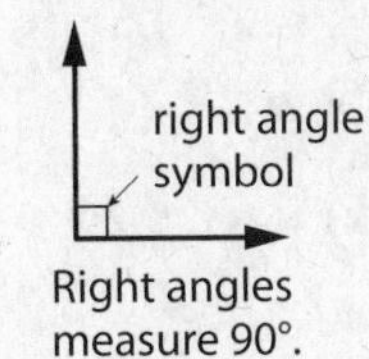

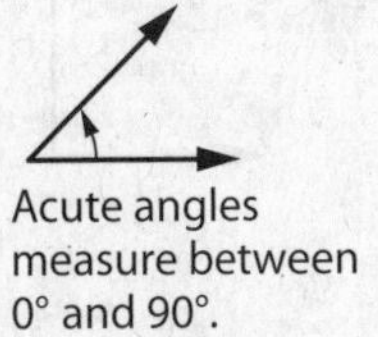

EXAMPLE 1 **Use a protractor to find the measure of the angle. Then classify the angle as *acute, obtuse, right,* or *straight*.**

To measure an angle, place the center of a protractor on the vertex of the angle. Place the zero mark of the scale along one side of the angle. Then read the angle measure where the other side of the angle crosses the scale.

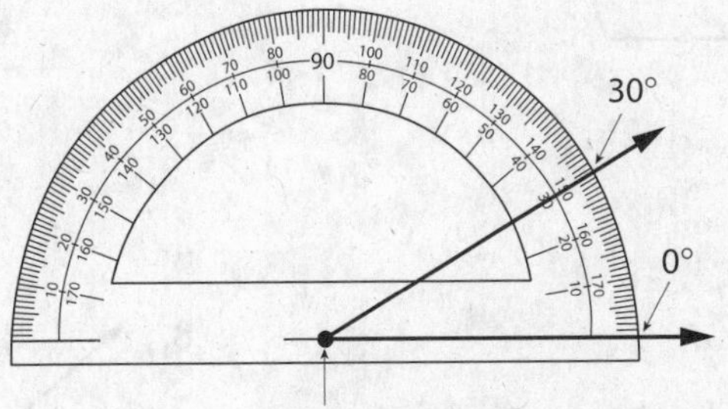

Align the center of the protractor.
This angle measures 30°.

The angle measures 30°. It is an acute angle.

Two angles are **complementary** if the sum of their measures is 90°.

Two angles are **supplementary** if the sum of their measures is 180°.

EXAMPLE 2 **ALGEBRA Angles *A* and *B* are complementary. If $m\angle A = 25°$, what is the measure of $\angle B$?**

$m\angle A + m\angle B = 90°$	Complementary angles
$25° + m\angle B = 90°$	Replace $m\angle A$ with 25°.
$25° + m\angle B - 25° = 90° - 25°$	Subtract 25° from each side.
$m\angle B = 65°$	

So, $m\angle B = 65°$. Since $25° + 65° = 90°$, the answer is correct.

EXERCISES

Use a protractor to find the measure of each angle. Then classify each angle as *acute, obtuse, right,* or *straight*.

1. 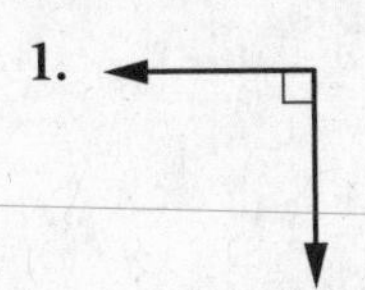

2. 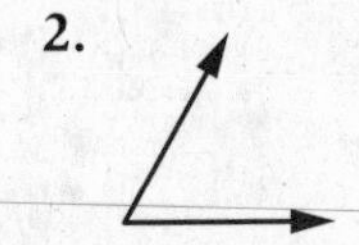

3. 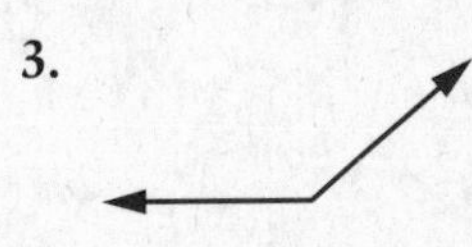

4. **ALGEBRA** Angles *K* and *L* are supplementary. If $m\angle L = 80°$, what is $m\angle K$?

5. **ALGEBRA** If $m\angle C = 40°$ and $\angle C$ and $\angle D$ are complementary, what is $m\angle D$?

Name________________________ Date________ Period________

Math Skills Study Guide

Angles

Use a protractor to find the measure of each angle. Then classify each angle as *acute*, *obtuse*, *right*, or *straight*.

1. 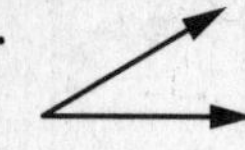

2. 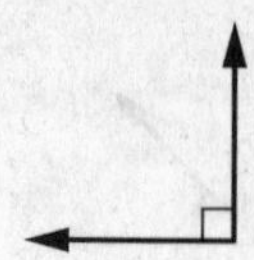

3. 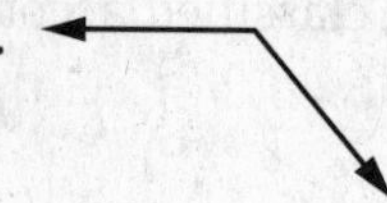

4. 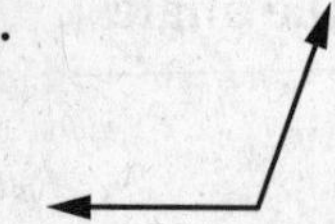

5. 

6. 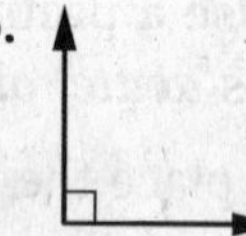

7. 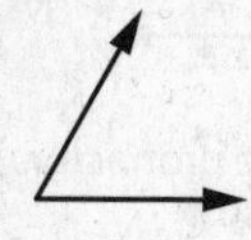

8. 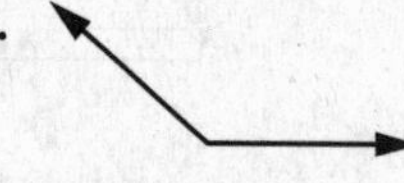

9. 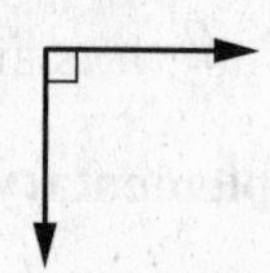

10. 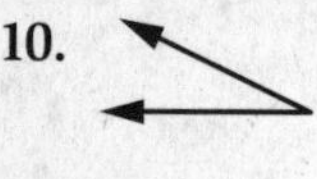

11. 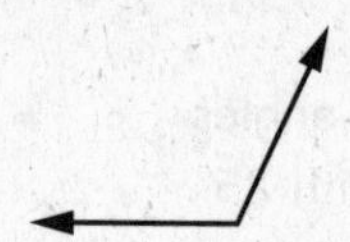

12. 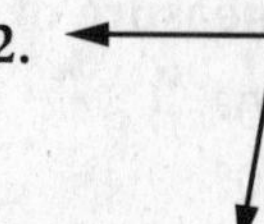

13. ALGEBRA If $m\angle K = 60°$ and $\angle J$ and $\angle K$ are complementary, what is $m\angle J$?

14. ALGEBRA Angles A and B are supplementary. What is $m\angle B$ if $m\angle A = 120°$?

Name ______________________ Date ________ Period ________

Math Skills Study Guide

Angle Relationships

Angles that have the same measure are called **congruent angles**. Two angles are **supplementary** if the sum of their measures is 180°. Two angles are **complementary** if the sum of their measures is 90°. When two lines intersect, they form two pairs of opposite angles called **vertical angles**, which are always congruent.

EXAMPLES **Classify each pair of angles as *complementary, supplementary,* or *neither.***

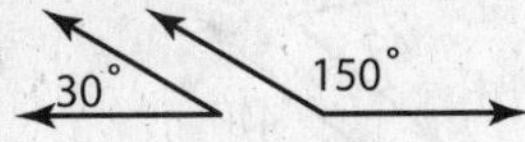

$30° + 150° = 180°$
The angles are supplementary.

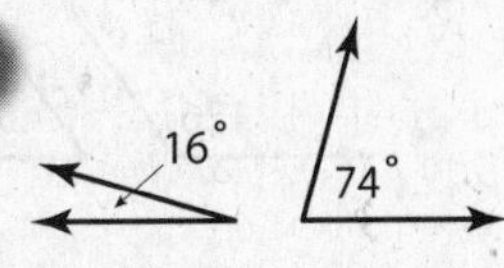

$16° + 74° = 90°$
The angles are supplementary.

EXAMPLE 3 **Find the value of *x* in the figure below.**

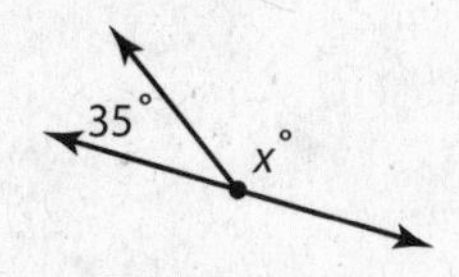

The two angles are supplementary, so the sum of their measures is 180°.

$x + 35 = 180$	Write the equation.
$\underline{\ \ -35 \quad -35}$	Subtract 35 from each side.
$x \quad\quad = 145$	Simplify.

So, the angle is 145°

EXERCISES

Classify each pair of angles as *complementary, supplementary,* or *neither.*

1.

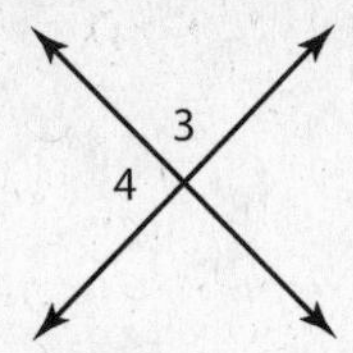

2.

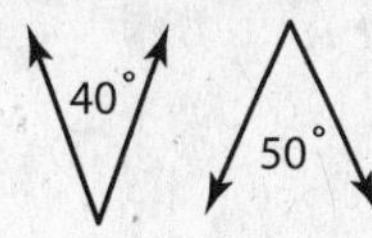

3.

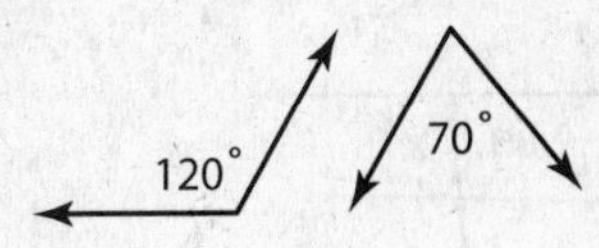

Find the value of *x* in each figure.

4.

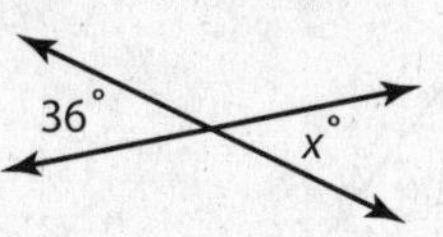

5.

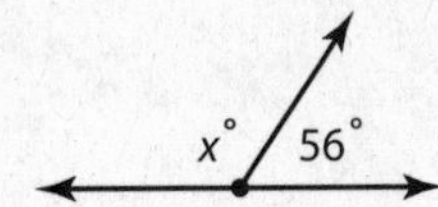

6.

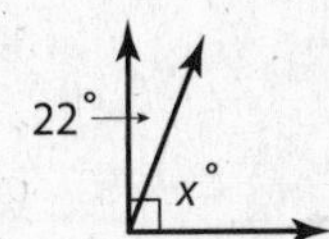

Name____________________________ Date________ Period________

Math Skills Study Guide

Angle Relationships

Classify each pair of angles as *complementary, supplementary,* or *neither.*

1.

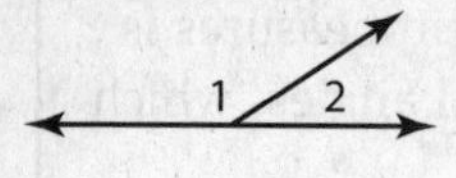

2.

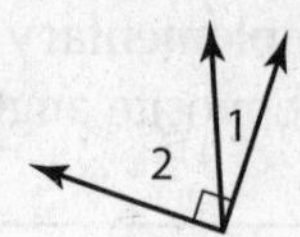

3.

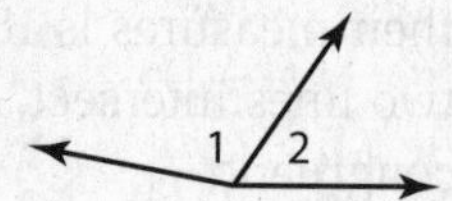

4.

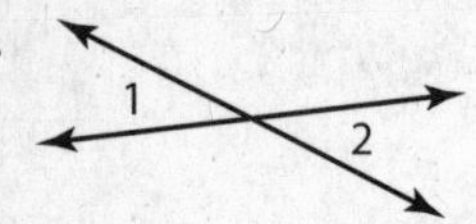

5.

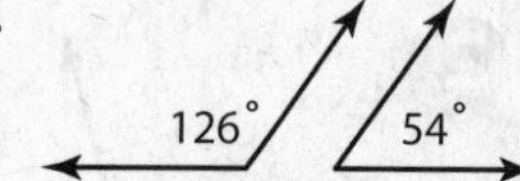

6.

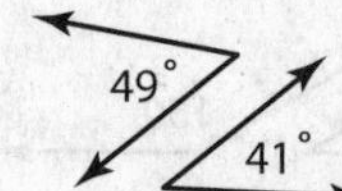

7.

8.

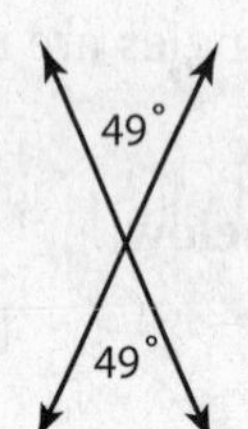

9.

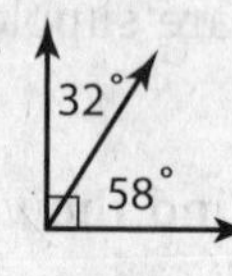

Find the value of *x* in each figure.

10.

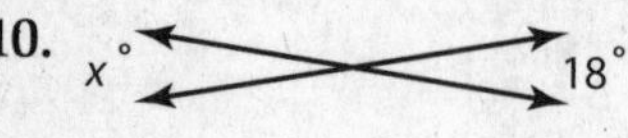

11.

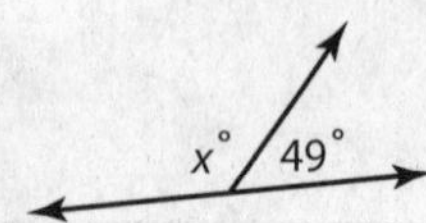

12.

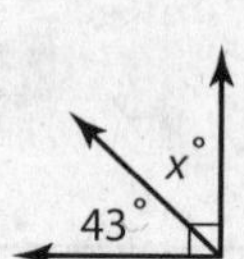

13.

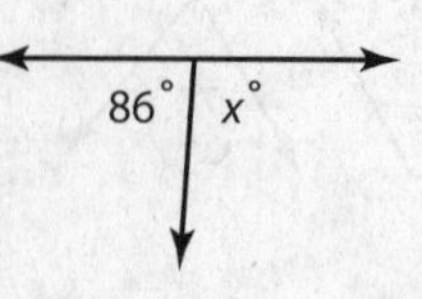

14.

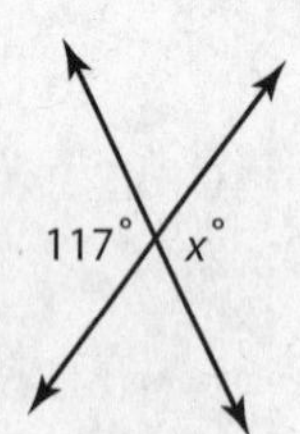

15.

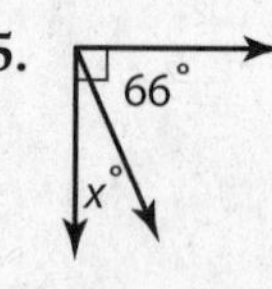

16.

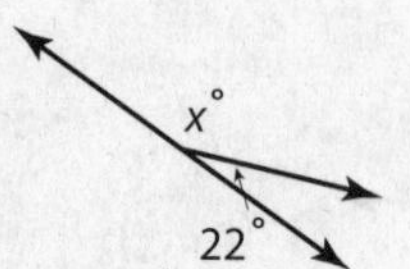

17.

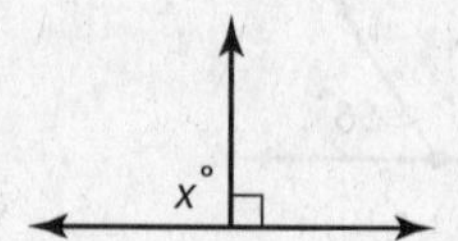

18.

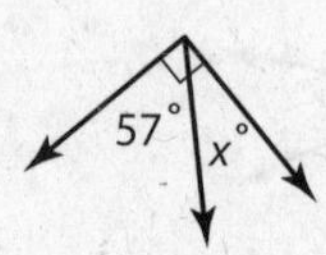

Name ______________________ Date ________ Period ________

Math Skills Study Guide

Bisectors

To **bisect** something means to separate it into two equal parts. You can use a straightedge and a compass to bisect line segments and angles.

EXAMPLE 1 **Use a straightedge and compass to bisect $\overline{BC}$.**

Draw $\overline{BC}$.

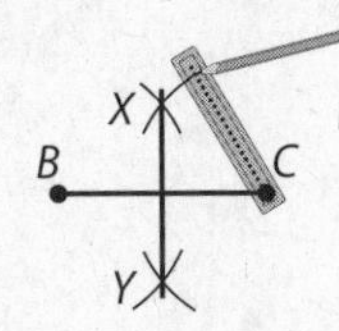

Place the compass at point B. Set the compass greater than half the length of $\overline{BC}$. Draw two arcs as shown. Use the same setting to place the compass point at C and draw another pair of arcs. Label the intersections X and Y.

Use a straightedge to align the intersections and draw a segment that intersects $\overline{BC}$. Label the intersection point M.

$\overline{XY}$ bisects $\overline{BC}$.

EXAMPLE 2 **Use a straightedge and compass to bisect $\angle ABC$.**

Draw $\angle ABC$.

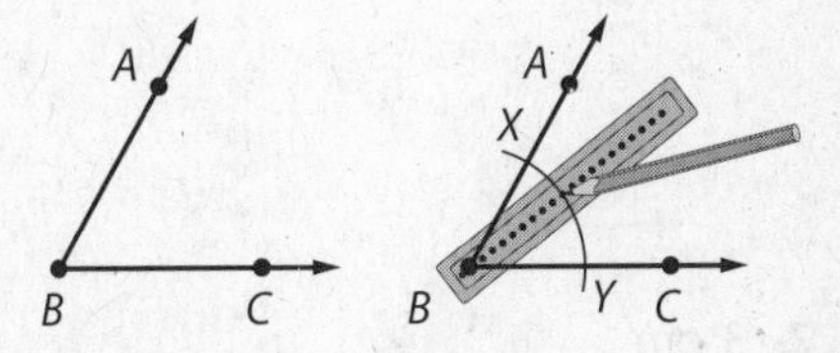

Place the compass at point B and draw an arc that intersects both sides of the angle. Label the intersections X and Y.

With the compass at point X, draw an arc as shown. Use the same setting to place the compass point at Y and draw another arc. Label the intersection Z.

Use a straightedge to draw $\overline{BZ}$. $\overline{BZ}$ bisects $\angle ABC$.

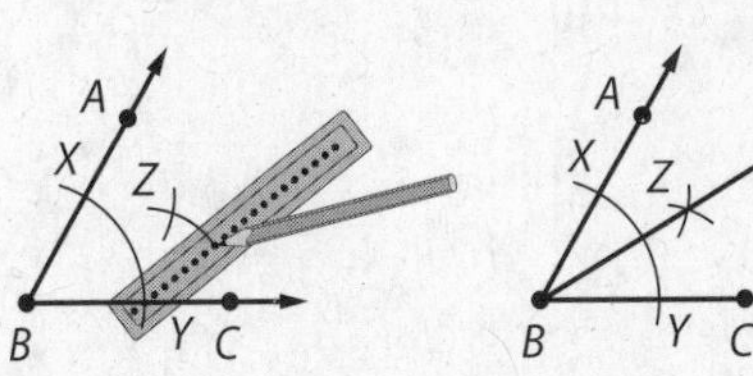

EXERCISES

Draw each line segment or angle having the given measurement. Then use a straightedge and a compass to bisect the line segment or angle.

1. 2 in.

2. 45°

3. 4 cm

4. 150°

Name________________________ Date______ Period________

Math Skills Study Guide

Bisectors

Draw each line segment or angle having the given measurement. Then use a straightedge and a compass to bisect the line segment or angle.

1. 80°

2. 1 in.

3. 120°

4. 40°

5. 2 cm

6. 90°

7. 3 cm

8. 78°

9. 1.25 in.

10. 25°

11. 165°

12. 2.25 in.

Name ______________________ Date ________ Period ________

Math Skills Study Guide

Triangles

A **triangle** is a figure with three sides and three angles. The sum of the measures of the angles of a triangle is 180°. You can use this to find a missing angle measure in a triangle.

EXAMPLE 1 **Find the value of x in $\triangle ABC$.**

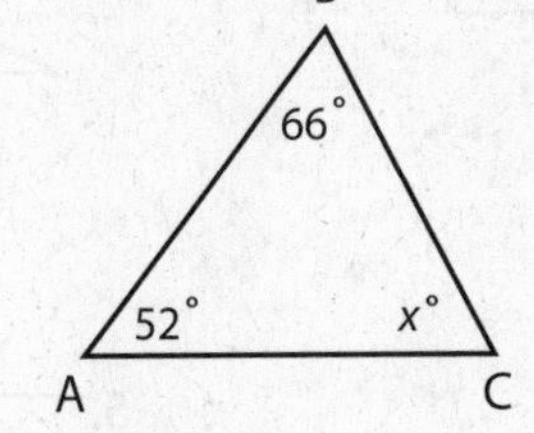

$x + 66 + 52 = 180$	The sum of the measures is 180.
$x + 118 = 180$	Simplify.
$-118 \quad -118$	Subtract 118 from each side.
$x = 62$	

The missing angle is 62°.

Triangles can be classified by the measures of their angles. An **acute triangle** has three acute angles. An **obtuse triangle** has one obtuse angle. A **right triangle** has one right angle.

Triangles can also be classified by the lengths of their sides. Sides that are the same length are **congruent segments** and are often marked by tick marks. In a **scalene triangle,** all sides have different lengths. An **isosceles** triangle has at least two congruent sides. An **equilateral triangle** has all three sides congruent.

EXAMPLE 2 **Classify the triangle by its angles and by its sides.**

The triangle has one obtuse angle and two sides the same length. So, it is an obtuse, isosceles triangle.

120°

EXERCISES

Find the missing measure in each triangle. Then classify the triangle as acute, right, or obtuse.

1.

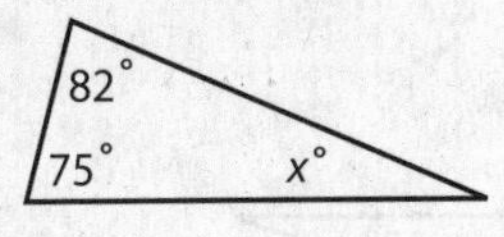

2.

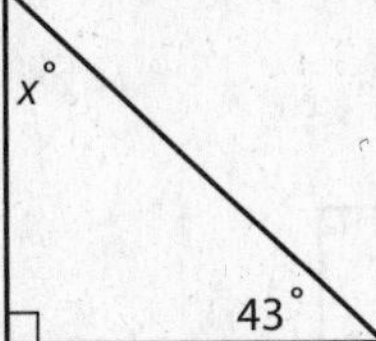

3.

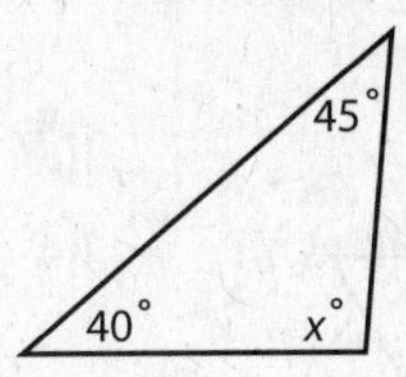

Classify each triangle by its angles and by its sides.

4.

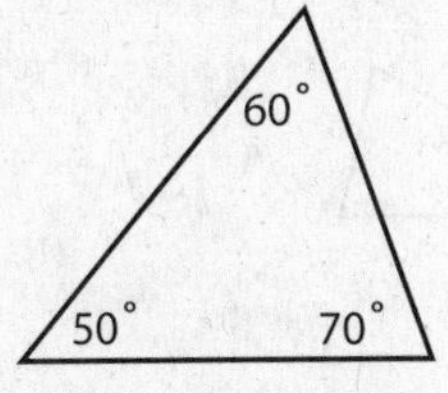

5.

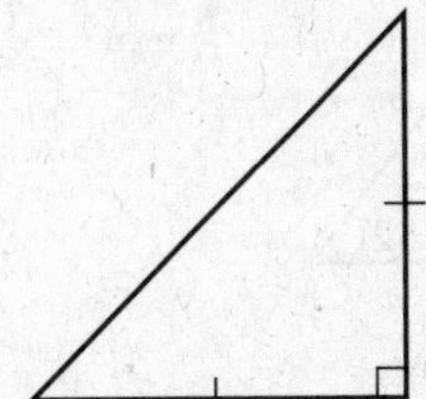

6.

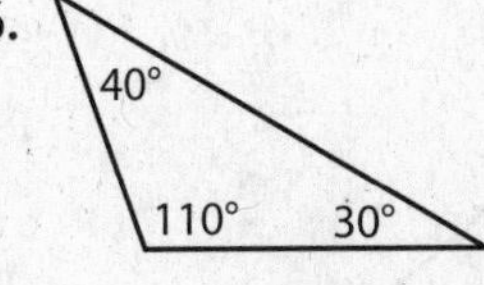

Name______________________________ Date________ Period________

Math Skills Study Guide

Triangles

Find the missing measure in each triangle. Then classify the triangle as *acute, right,* or *obtuse.*

1.

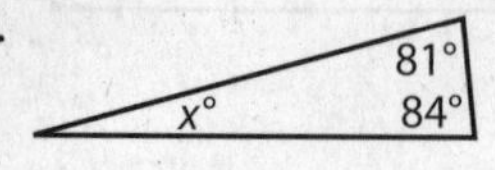

2.

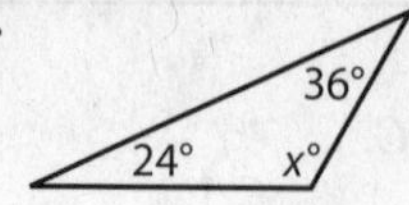

3.

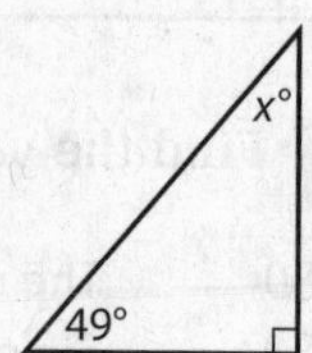

4.

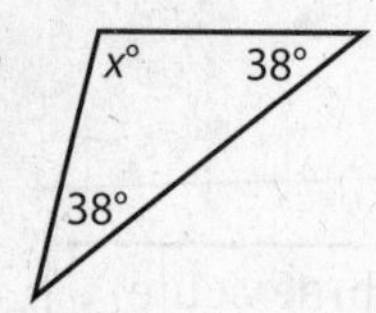

5.

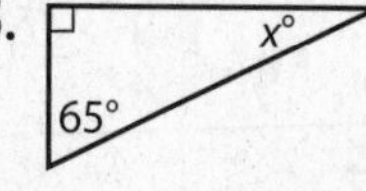

6.

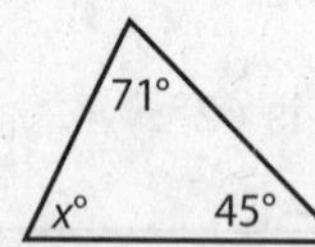

7.

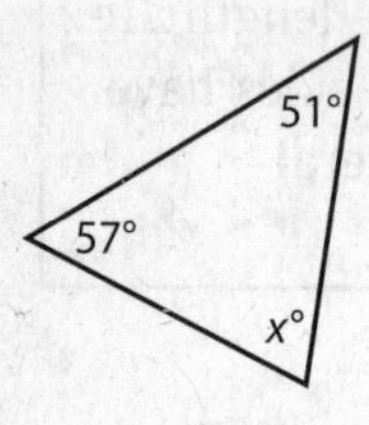

8.

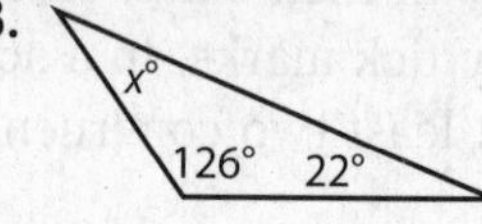

9.

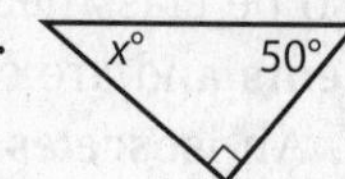

Classify each triangle by its angles and by its sides.

10.

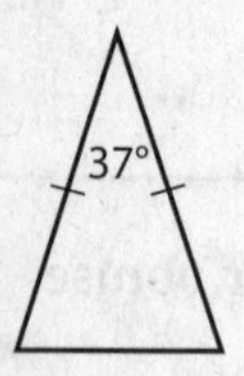

11.

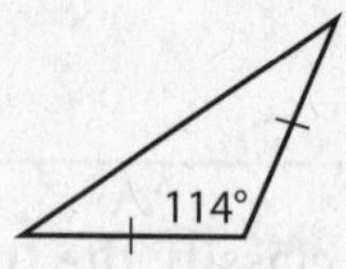

12.

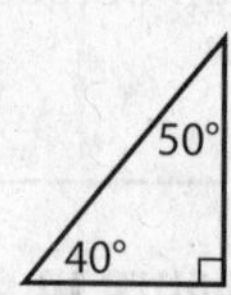

13.

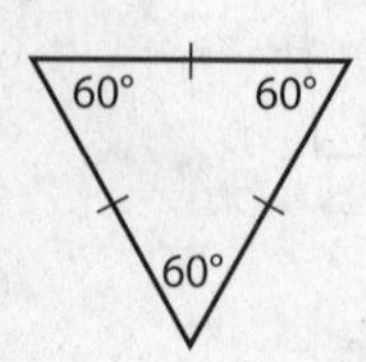

14.

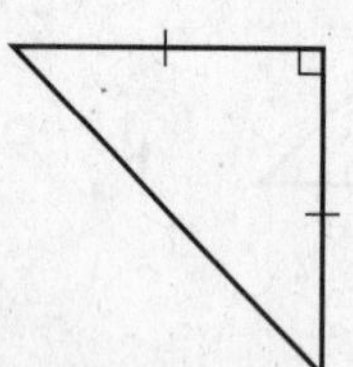

15.

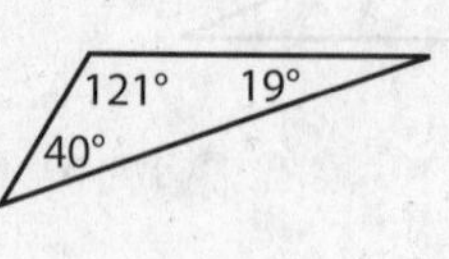

16.

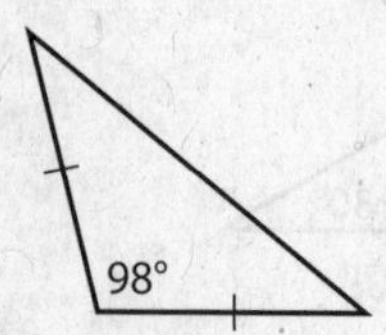

17.

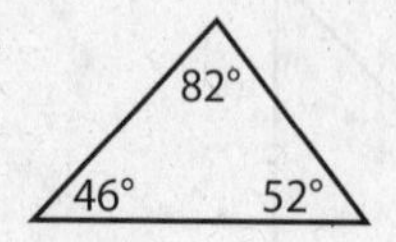

18.

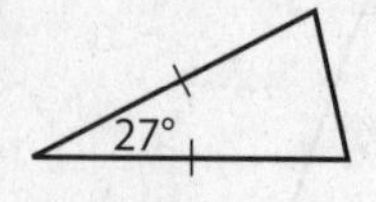

Math Skills Study Guide

Quadrilaterals

A **quadrilateral** is a closed figure with four sides and four vertices. The segments that make up a quadrilateral intersect only at their endpoints.

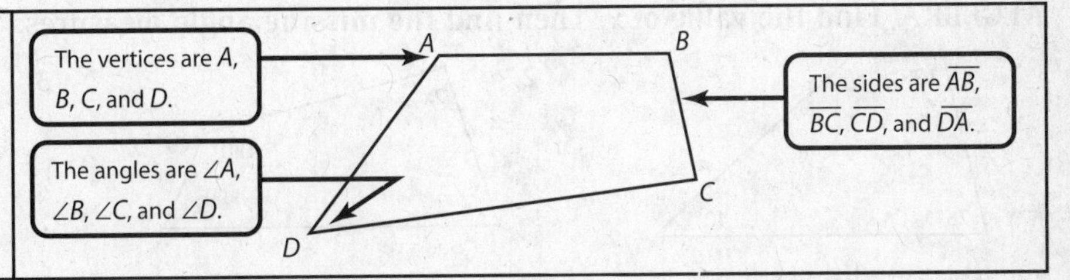

A quadrilateral can be seperated into two triangles. Since the sum of the measures of the angles of a triangle is 180°, the sum of the measures of the angles of a quadrilateral is 2(180°) or 360°.

EXAMPLE **ALGEBRA Find the value of *x*. Then find each missing angle measure.**

Words The sum of the measures of the angles is 360°.

Variable Let $m\angle A$, $m\angle B$, $m\angle C$, and $m\angle D$ represent the measures of the angles.

Equation

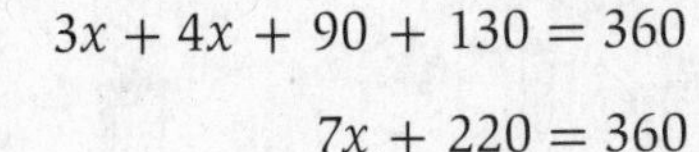

$m\angle A + m\angle B + m\angle C + m\angle D = 360$	Angles of a quadrilateral
$3x + 4x + 90 + 130 = 360$	Substitution
$7x + 220 = 360$	Combine like terms.
$7x + 220 - 220 = 360 - 220$	Subtract 220 from each side.
$7x = 140$	Simplify.
$x = 20$	Divide each side by 7.

The value of x is 20. So, $m\angle A = 3(20)$ or 60° and $m\angle B = 4(20)$ or 80°.

EXERCISES

ALGEBRA Find the value of *x*. Then find the missing angle measures.

1. 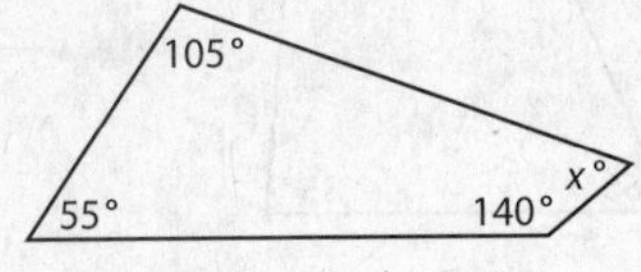

2. 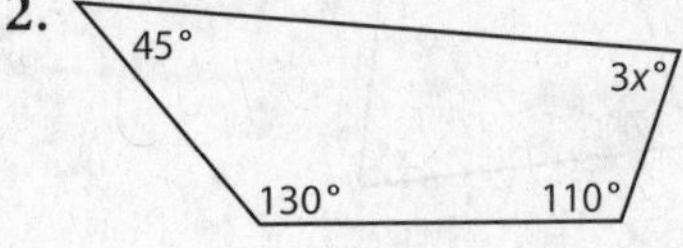

3. 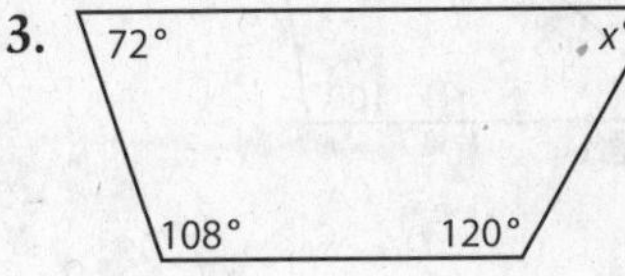

Name______________________________ Date__________ Period__________

Math Skills Study Guide

Quadrilaterals

ALGEBRA Find the value of x. Then find the missing angle measures.

1.

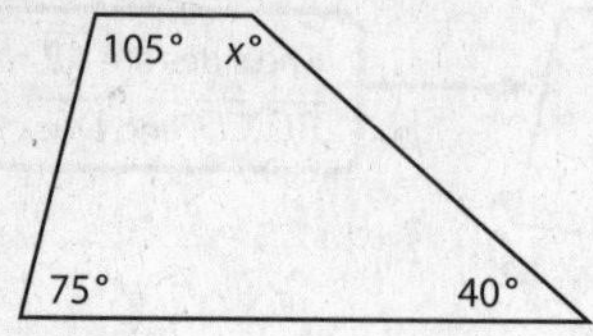

2.

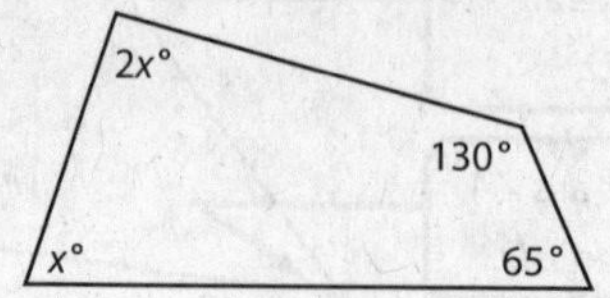

3.

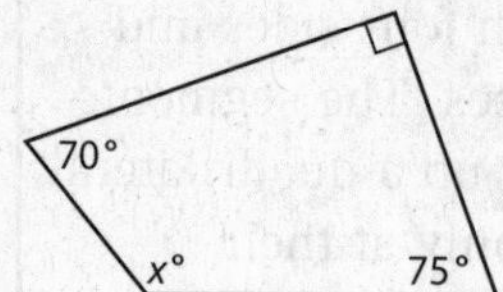

4.

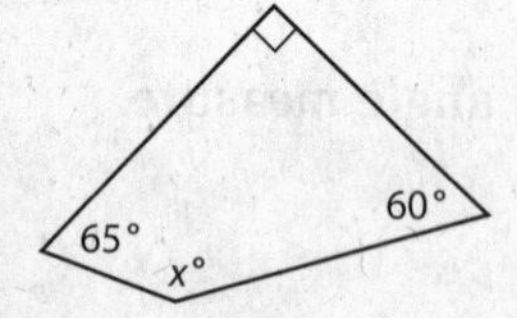

5.

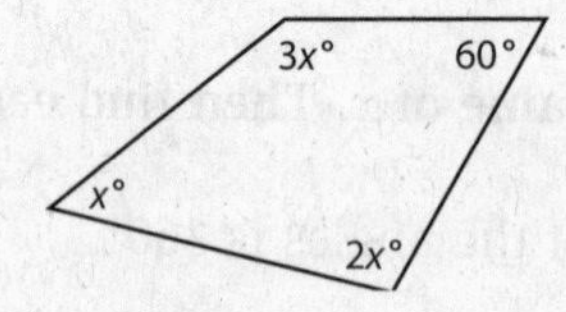

6.

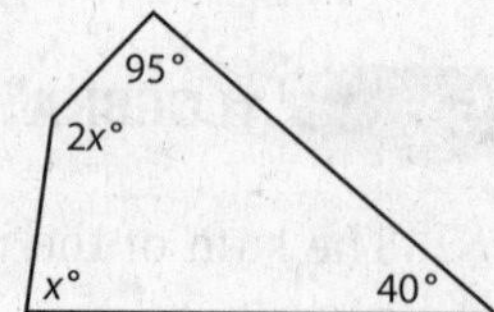

7.

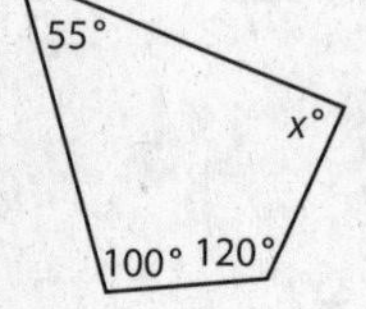

8.

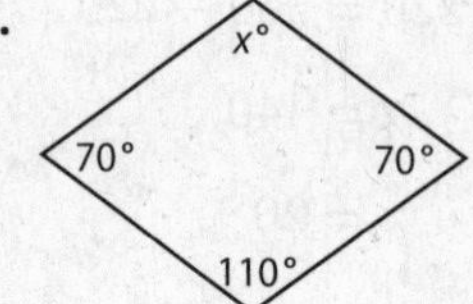

9.

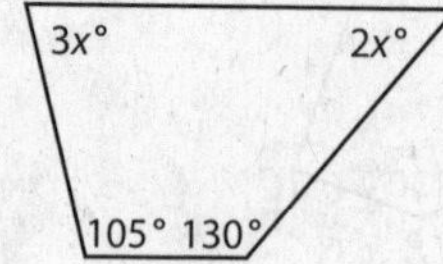

10.

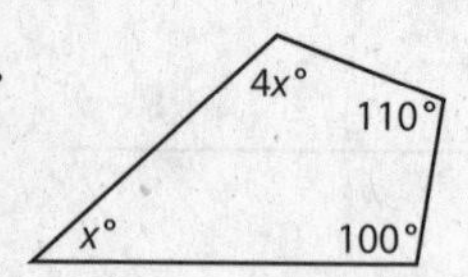

11.

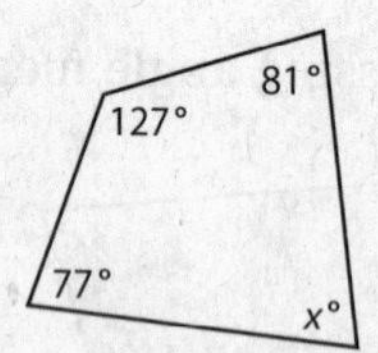

12.

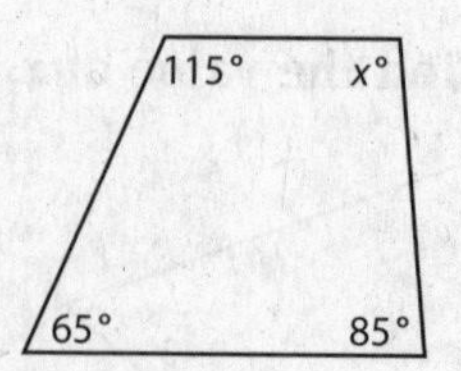

Name ______________________ Date ________ Period ________

Math Skills Study Guide

Lines of Symmetry

If a figure can be folded in half so that the two halves match exactly, the figure has **line symmetry**. The line that separates the figure into two matching halves is a **line of symmetry**. If a figure can be rotated less than 360° and look exactly as it did before being turned, the figure has **rotational symmetry**.

EXAMPLES **Draw all lines of symmetry for each figure.**

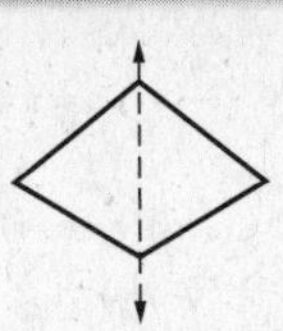

one line of symmetry

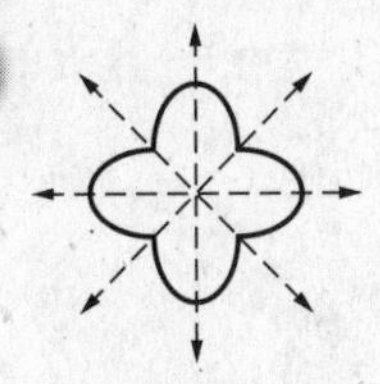

four lines of symmetry

no lines of symmetry

EXAMPLES **Tell whether each figure has rotational symmetry.**

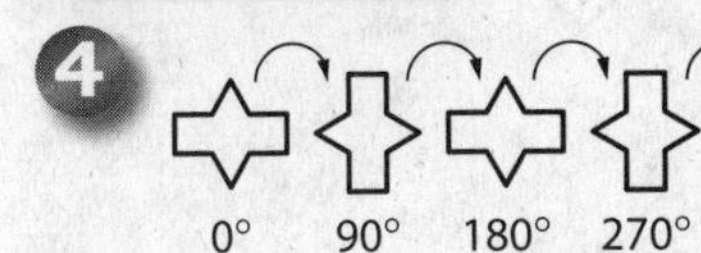

The figure appears as it did before being rotated after being rotated 180°. So, the figure has rotational symmetry.

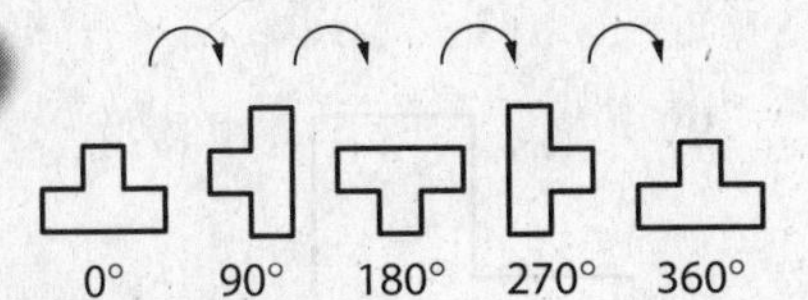

The figure looks the same only when rotated 360°. So, the figure does not have rotational symmetry.

EXERCISES

Draw all lines of symmetry for each figure.

1.

2.

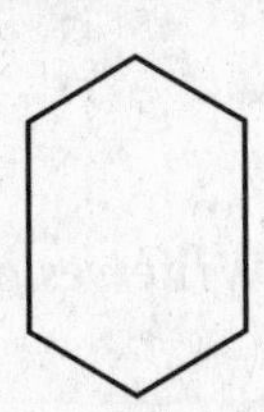

3.

Tell whether each figure has rotational symmetry. Write *yes* or *no*.

4.

5.

6.

Name ____________________ Date ________ Period ________

Math Skills Study Guide

Lines of Symmetry

Draw all lines of symmetry for each figure.

1.

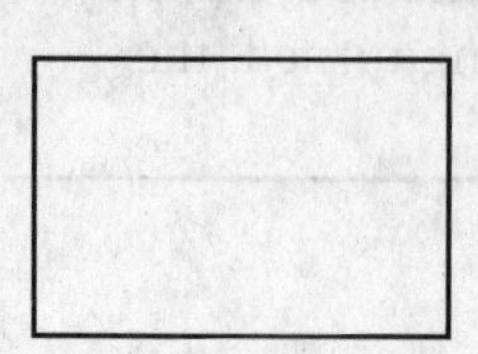

2.

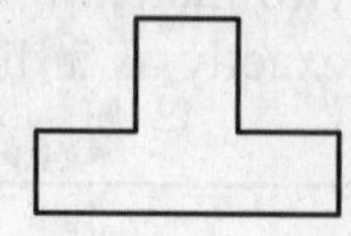

3.

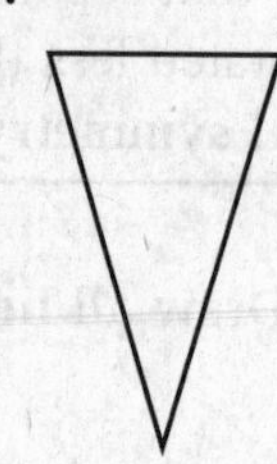

4.

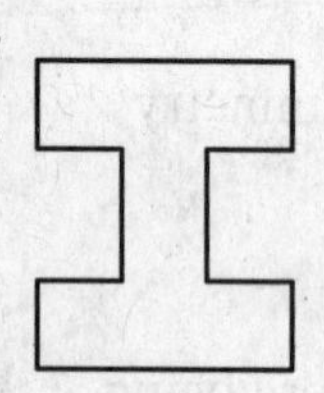

5.

6.

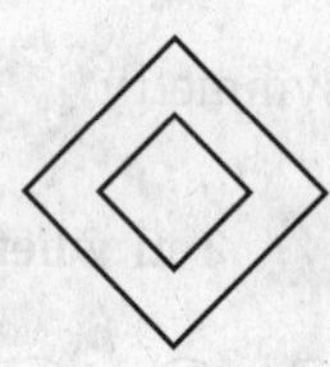

7.

8.

9.

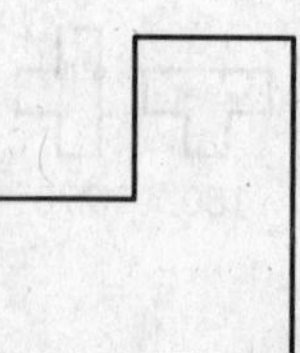

Tell whether each figure has rotational symmetry. Write *yes* or *no*.

10.

11.

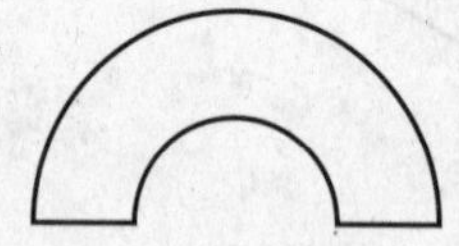

12.

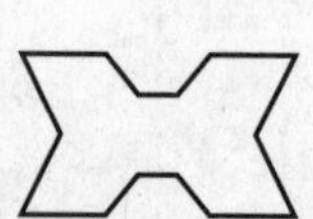

Name ______________________ Date ______ Period ______

Math Skills Study Guide

Circumference

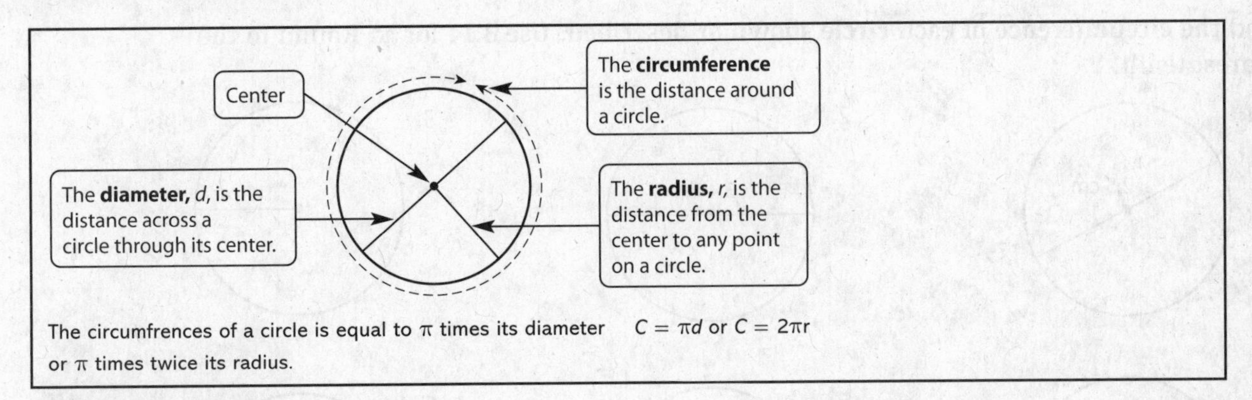

The circumfrences of a circle is equal to π times its diameter or π times twice its radius. $C = \pi d$ or $C = 2\pi r$

EXAMPLE 1 **Find the circumference of a circle whose diameter is 4.2 meters. Round to the nearest tenth.**

$C = \pi d$	Write the formula.
$\approx 3.14 \times 4.2$	Replace π with 3.14 and *d* with 4.2.
≈ 13.188	Multiply.
≈ 13.2	Round to the nearest tenth.

The circumference of the circle is about 13.2 meters.

EXAMPLE 2 **Find the circumference of a circle whose radius is 13 inches. Round to the nearest tenth.**

$C = 2\pi r$	Write the formula.
$\approx 2 \times 3.14 \times 13$	Replace π with 3.14 and *r* with 13.
≈ 81.64	Multiply.
≈ 81.6	Round to the nearest tenth.

The circumference of the circle is about 81.6 inches.

EXERCISES

Find the circumference of each circle shown or described. Round to the nearest tenth.

1.

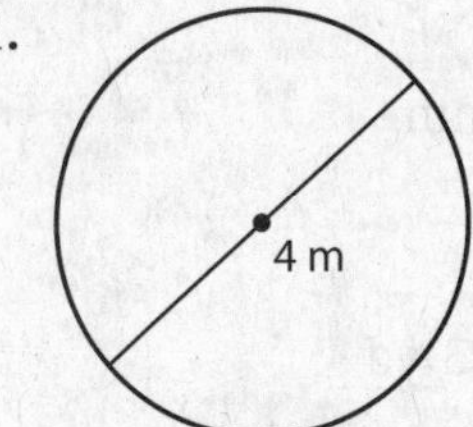

2.

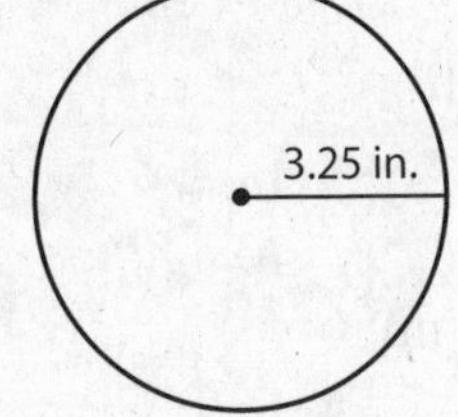

3.

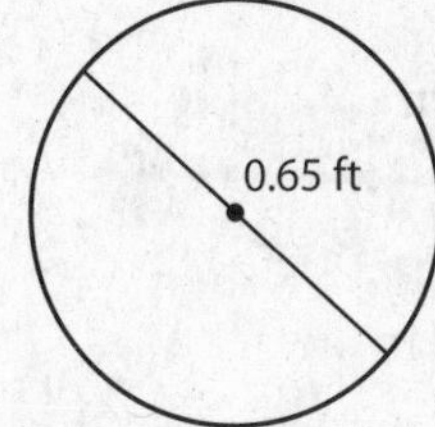

4. The radius of a circle measures 16 miles. What is the measure of its circumference to the nearest tenth?

5. Find the circumference of a circle whose diameter is 12.5 yards to the nearest tenth.

6. What is the circumference of a circle with a radius of 2.05 inches to the nearest tenth?

Name______________________________ Date________ Period________

Math Skills Study Guide

Circumference

Find the circumference of each circle shown or described. Use 3.14 for π. Round to the nearest tenth.

1.

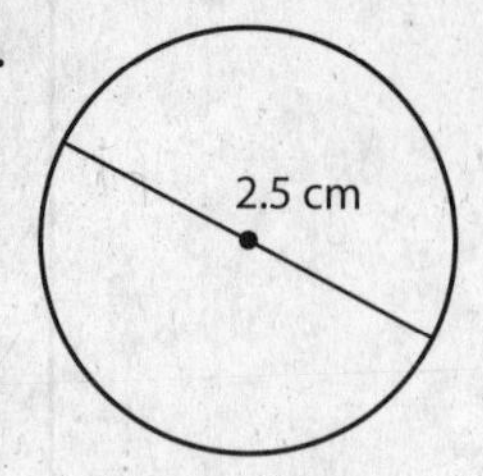

2.

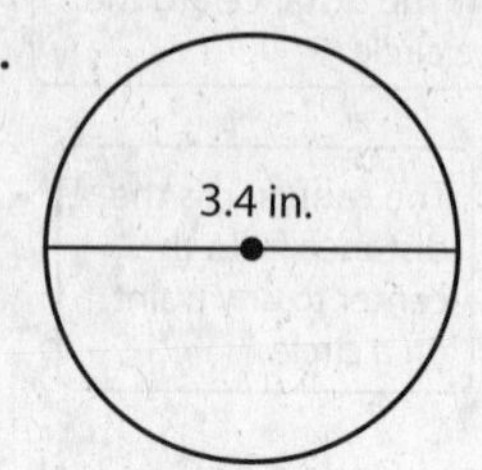

3.

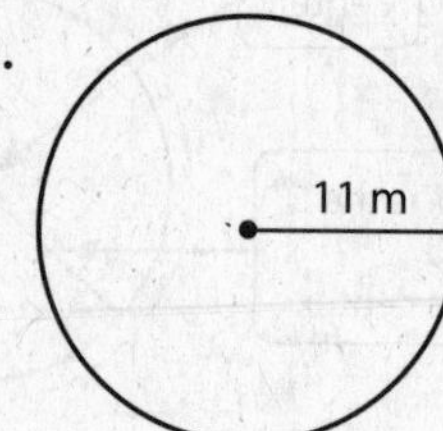

4.

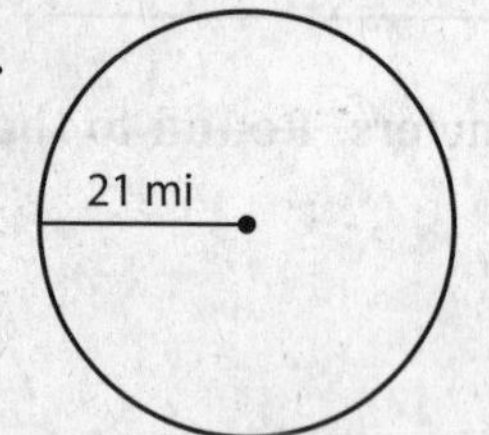

5.

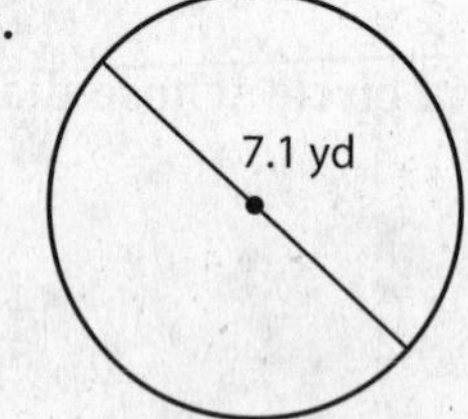

6.

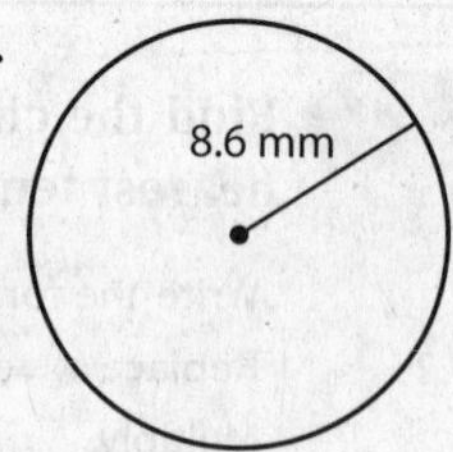

7.

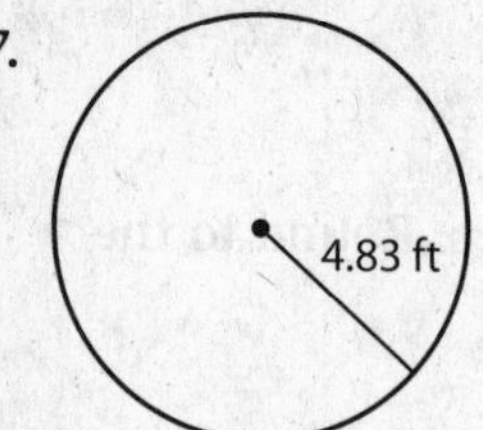

8.

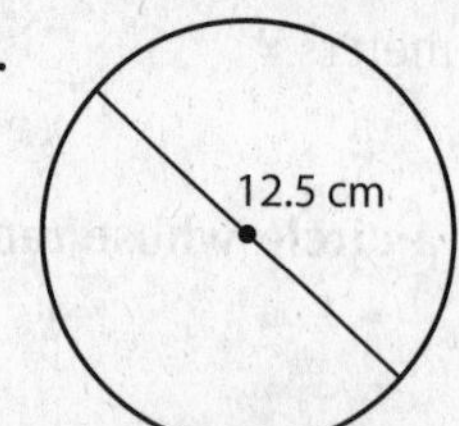

9.

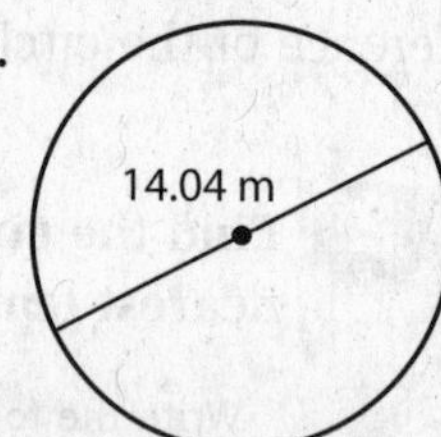

10.

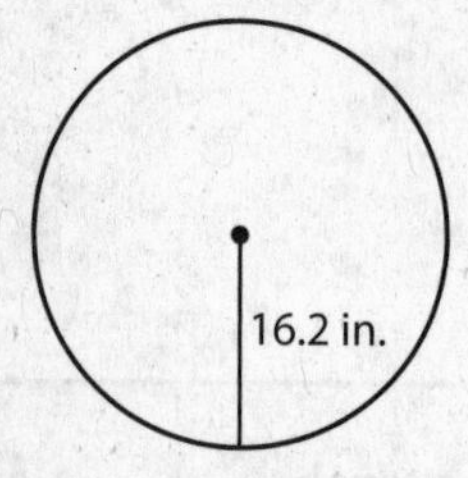

11.

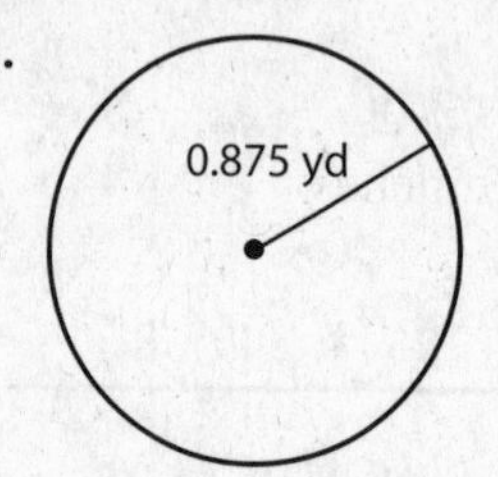

12.

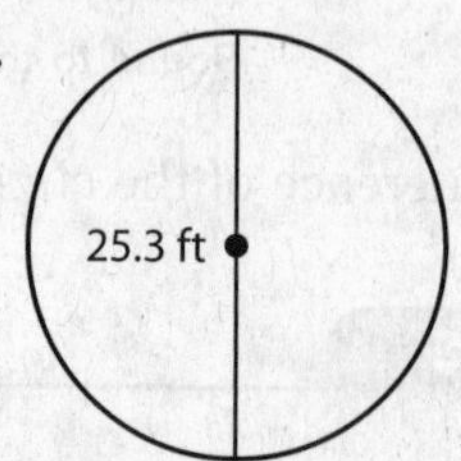

13. $r = 13$ cm

14. $d = 4.1$ ft

15. $r = 22$ mm

16. $d = 1.25$ in.

17. $r = 10.6$ mi

18. $d = 14.23$ yd

Name ______________________ Date ________ Period ________

Math Skills Study Guide

Geometry: Perimeter

The distance around a geometric figure is called the **perimeter.**
To find the perimeter of any geometric figure, add the measures of its sides.
The perimeter of a rectangle is twice the length ℓ plus twice the width w.

$$P = 2\ell + 2w$$

EXAMPLE 1 **Find the perimeter of the figure at the right.**

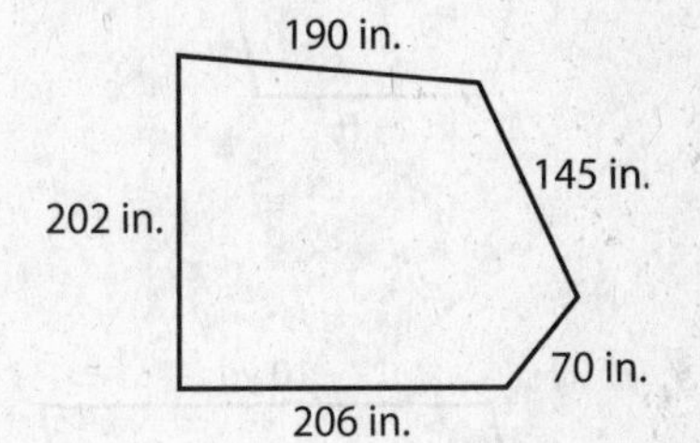

$P = 145 + 70 + 206 + 202 + 190$
$= 813$

The perimeter is 813 inches.

EXERCISES

Find the perimeter of each figure.

1. 7 cm, 9 cm, 15 cm

2. $7\frac{7}{8}$ ft, $3\frac{3}{4}$ ft, $12\frac{1}{8}$ ft, 10 ft, $6\frac{1}{4}$ ft, 20 ft

Find the perimeter of each rectangle.

3. 4 ft, 9 ft

4. 11 in., 3 in.

5. $\ell = 8$ ft, $w = 5$ ft

6. $\ell = 3.5$ m, $w = 2$ m

7. $\ell = 8$ yd, $w = 4\frac{1}{3}$ yd

8. $\ell = 29$ cm, $w = 7.3$ cm

Name______________________ Date________ Period________

Math Skills Study Guide

Geometry: Perimeter

Find the perimeter of each figure.

1.

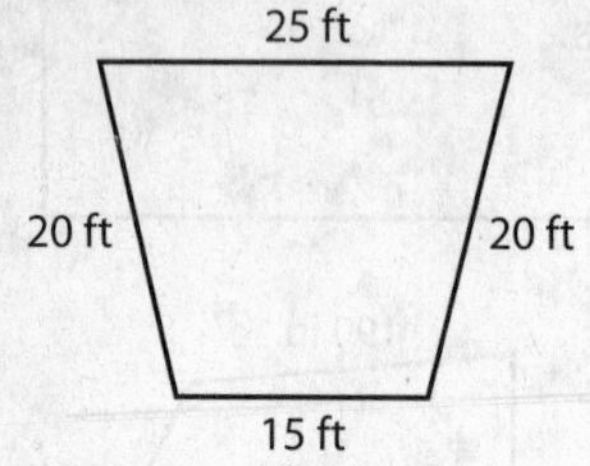

2.

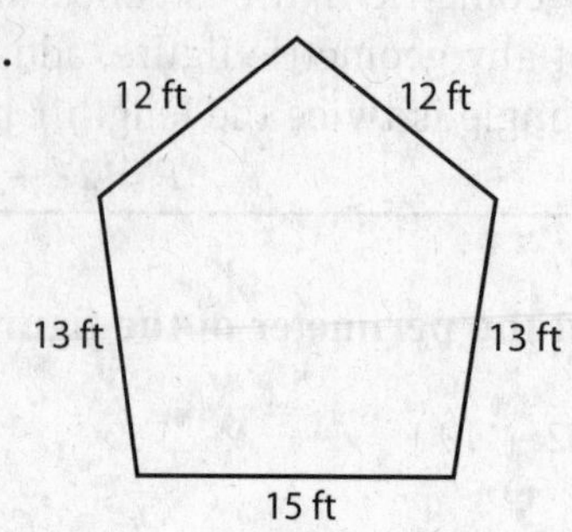

3.

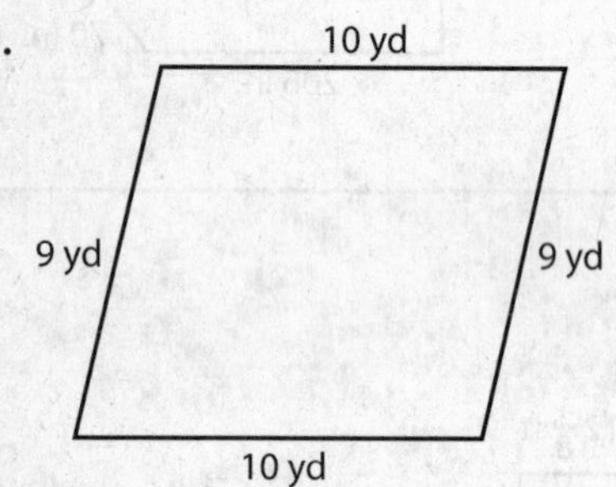

4.

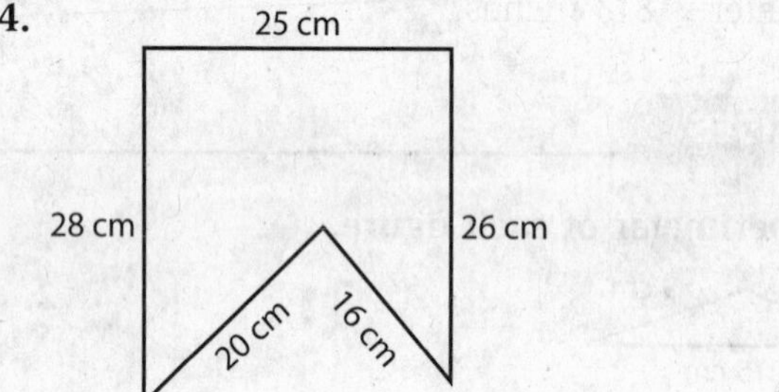

Find the perimeter of each rectangle.

5.

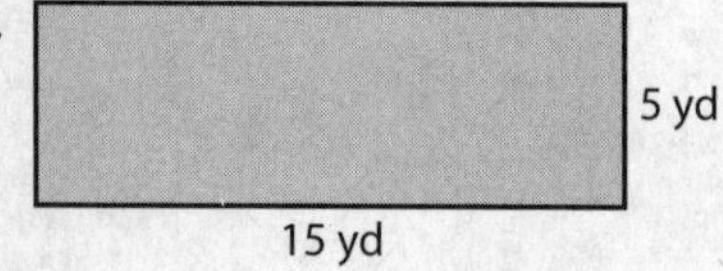

6.

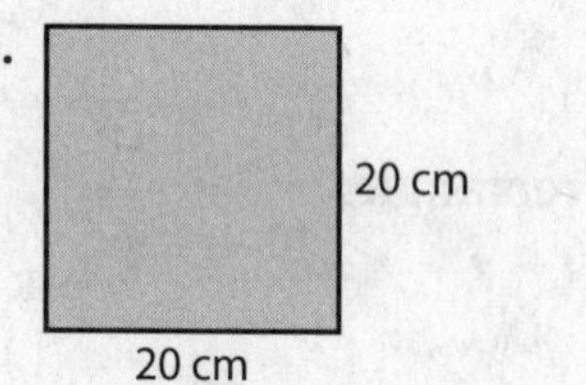

7.

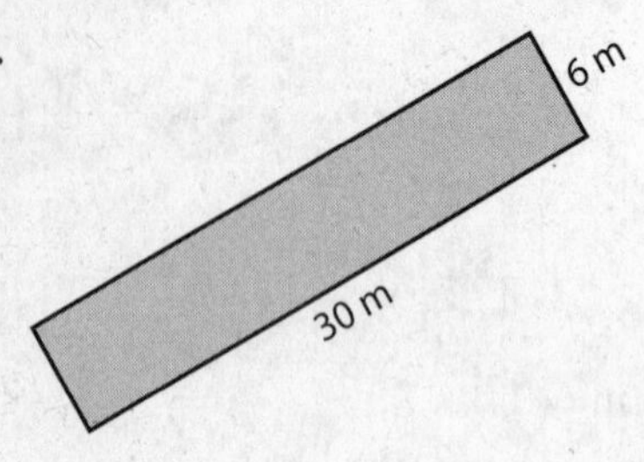

8.

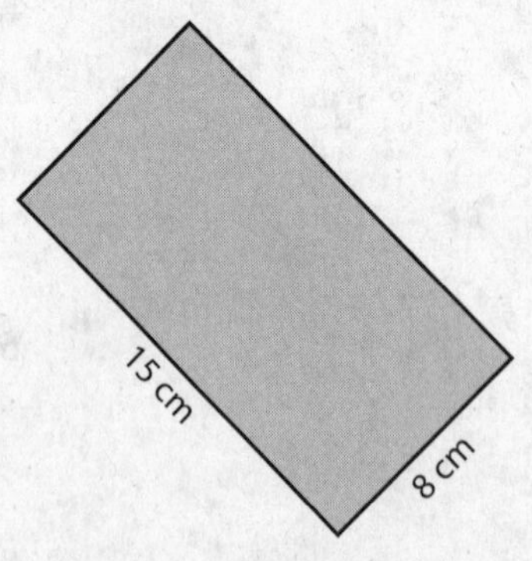

9. $\ell = 6$ yd, $w = 4$ yd

10. $\ell = 8.2$ m, $w = 7.1$ m

11. $\ell = 50$ in., $w = 10$ in.

12. $\ell = 10$ cm, $w = 4\frac{1}{2}$ cm

13. $\ell = 4.5$ ft, $w = 3$ ft

14. $\ell = 7\frac{1}{2}$ mm, $w = 6\frac{3}{8}$ mm

Name ______________________ Date ________ Period ________

Math Skills Study Guide

Geometry: Area of Rectangles

The **area** of a figure is the number of square units needed to cover its surface. You can use a formula to find the area of a rectangle. The formula for finding the area of a rectangle is $A = \ell \times w$. In this formula, A represents area, ℓ represents the length of the rectangle, and w represents the width of the rectangle.

EXAMPLE 1 **Find the area of a rectangle with length 8 feet and width 7 feet.**

$A = \ell \times w$ Area of a rectangle
$A = 8 \times 7$ Replace ℓ with 8 and w with 7.
$A = 56$

The area is 56 square feet.

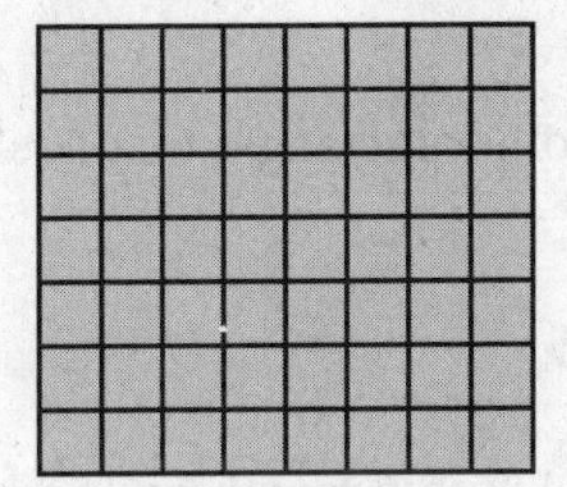

EXAMPLE 2 **Find the area of a rectangle with width 5 inches and length 6 inches.**

$A = \ell \times w$ Area of a rectangle
$A = 6 \times 5$ Replace ℓ with 6 and w with 5.
$A = 30$

The area is 30 square inches.

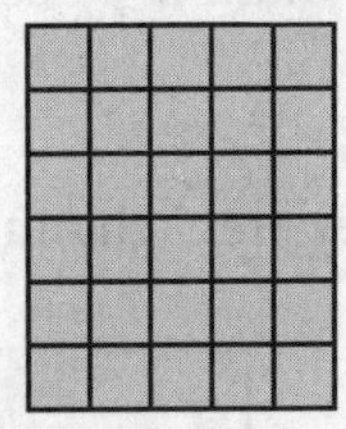

EXERCISES

Find the area of each rectangle.

1.

2. 5 ft, 8 ft

3. 7 cm, 3 cm

4. 6 yd, 5 yd

5. What is the area of a rectangle with a length of 10 meters and a width of 7 meters?

6. What is the area of a rectangle with a length of 35 inches and a width of 15 inches?

Name________________________________ Date________ Period________

Math Skills Study Guide

Geometry: Area of Rectangles

Complete each problem.

1. Give the formula for finding the area of a rectangle.

2. Draw and label a rectangle that has an area of 18 square units.

3. Give the dimensions of another rectangle that has the same area as the one in Exercise 2.

4. Find the area of a rectangle with a length of 3 miles and a width of 7 miles.

5. Find the area of a rectangle with a width of 54 centimeters and a length of 12 centimeters.

Find the area of each figure. All angles are right angles.

6.

9 in.
6 in.

7.

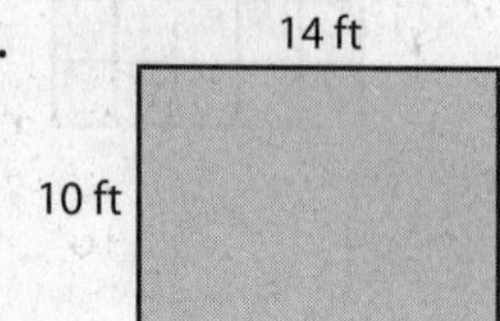

8.

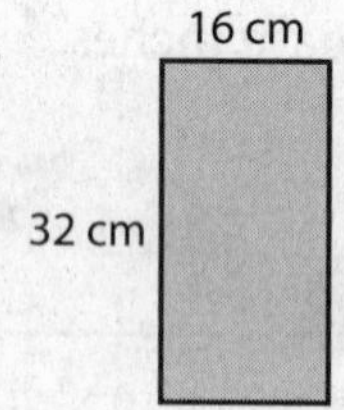

9.

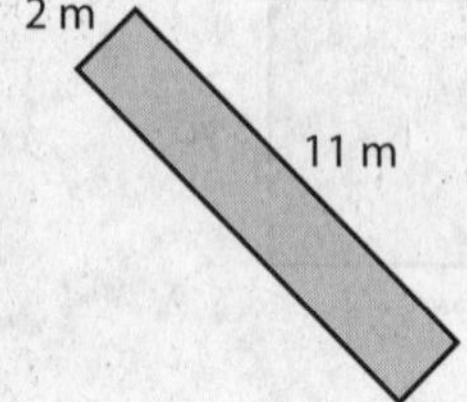

10.

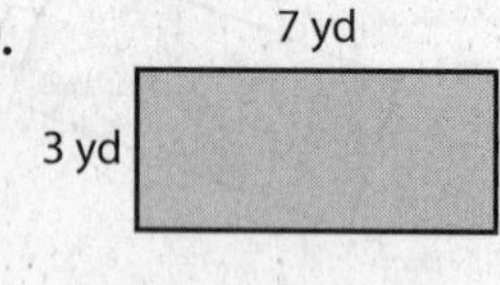

11.

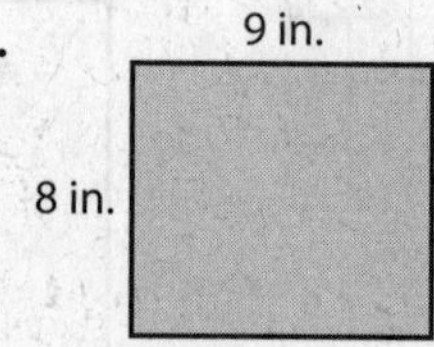

12.

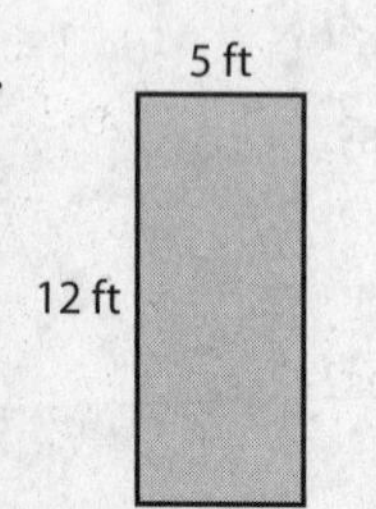

13.

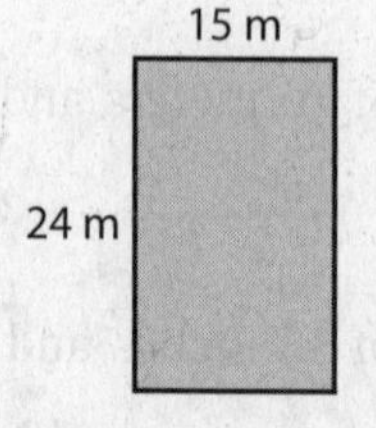

14.

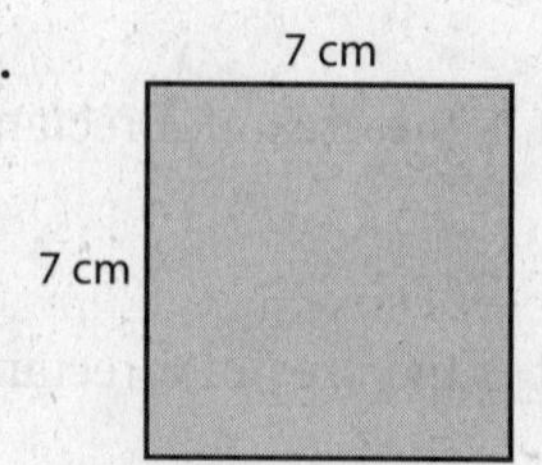

Name ______________________________ Date ________ Period ________

Math Skills Study Guide

Area of Parallelograms

The area A of a parallelogram is the product of any base b and its height h.

Symbols $A = bh$

Model

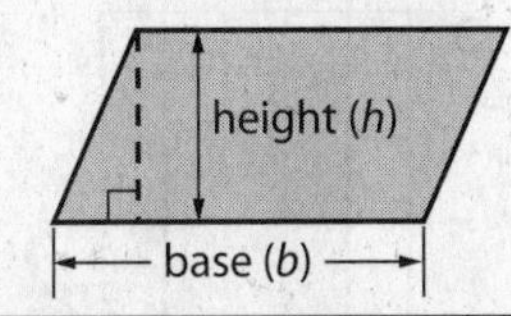

EXAMPLES Find the area of each parallelogram.

1

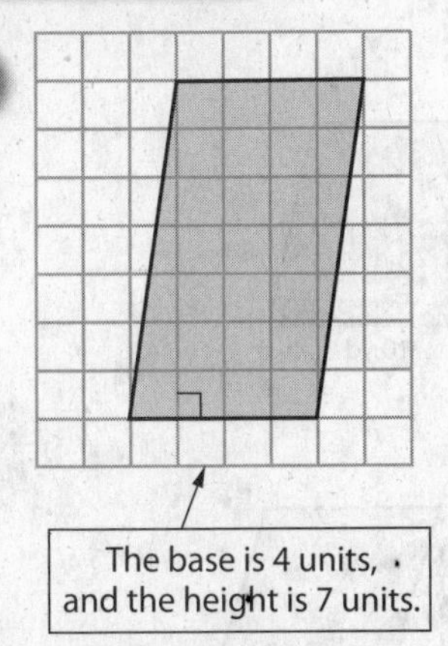

$A = bh$
$A = 4 \times 7$
$A = 28$
The area is 28 square units or 28 units^2.

The base is 4 units, and the height is 7 units.

2

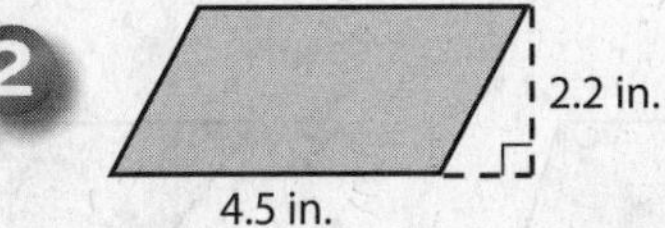

$A = bh$
$A = 4.5 \times 2.2$
$A = 9.9$
The area is 9.9 square inches or 9.9 in^2.

EXERCISES

Find the area of each parallelogram. Round decimals to the nearest tenth.

1.

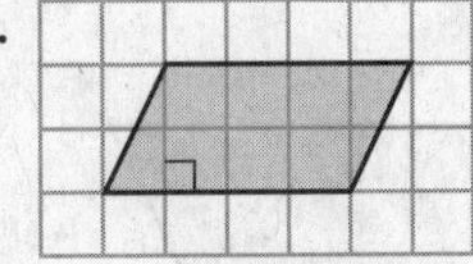

2.

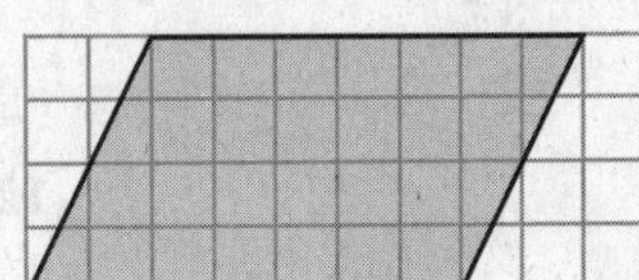

3.

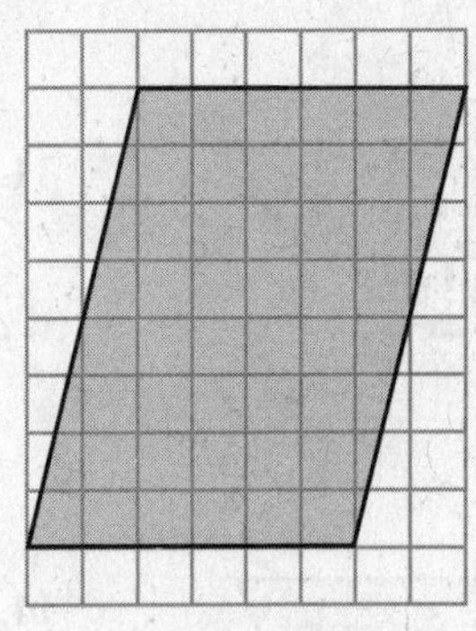

4.

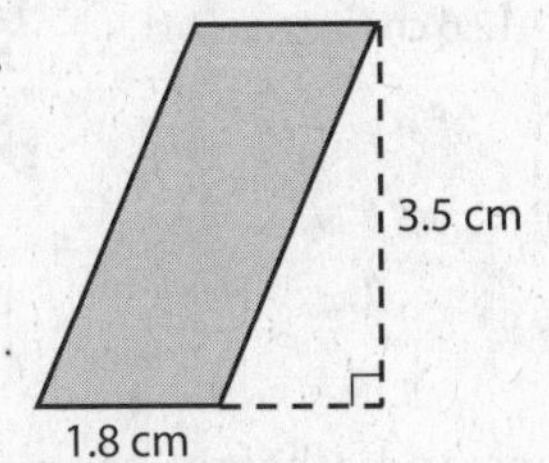

5.

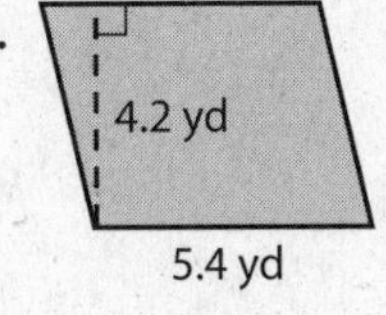

6.

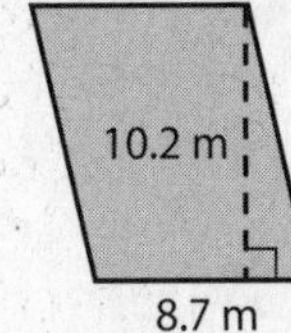

Name______________________________ Date________ Period________

Math Skills Study Guide

Area of Parallelograms

Find the area of each parallelogram. Round decimals to the nearest tenth.

1.

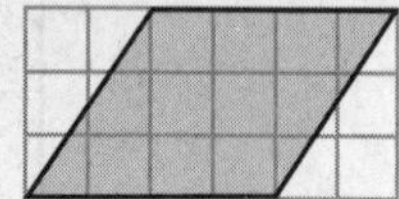

2.

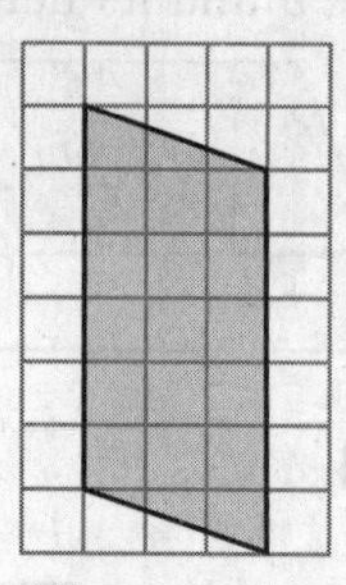

3.

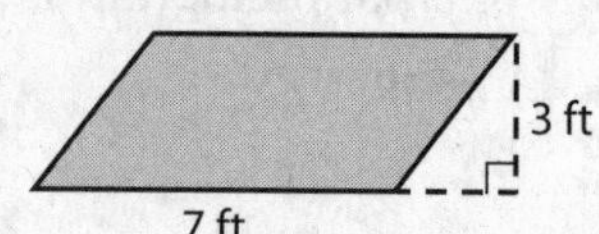

4.

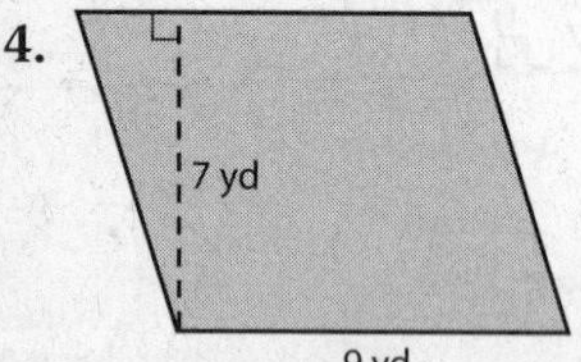

5.

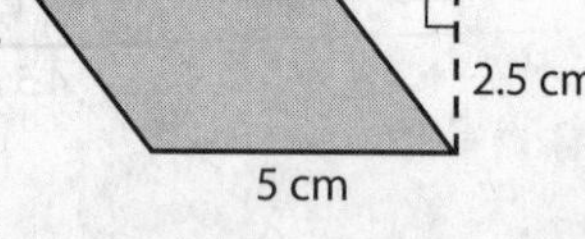

6.

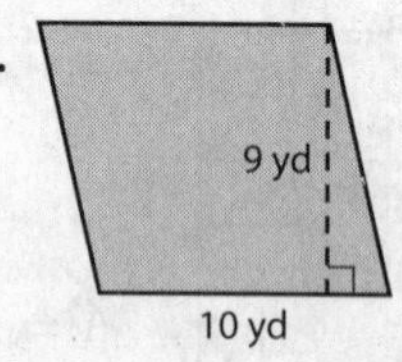

7.

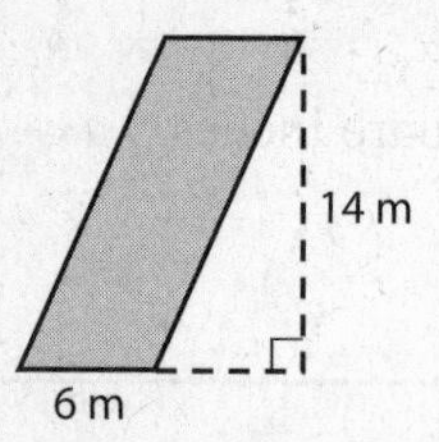

8.

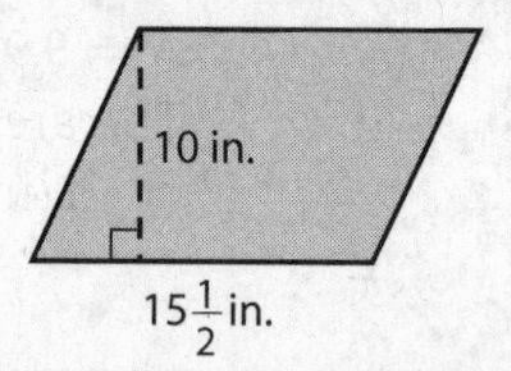

9.

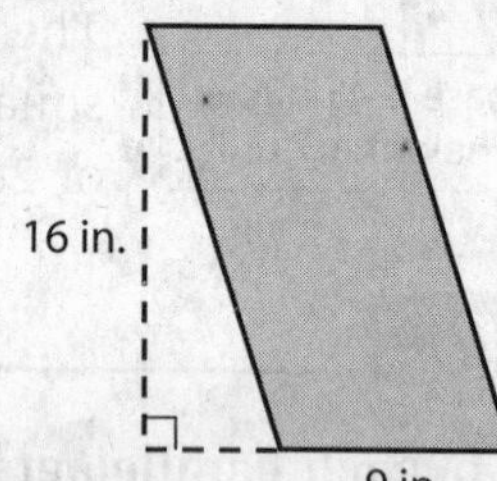

10.

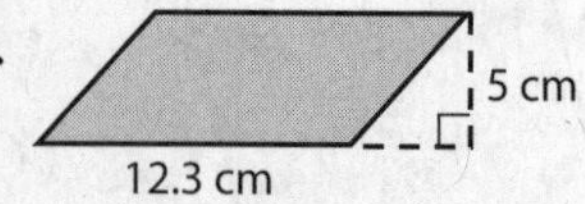

11.

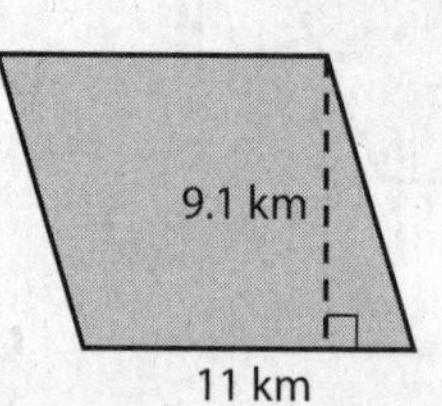

12.

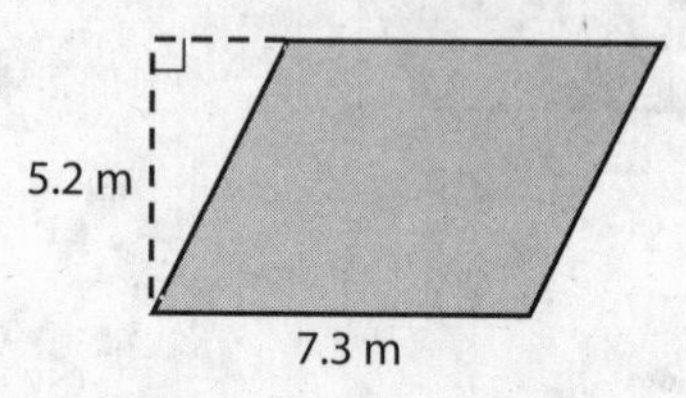

13.

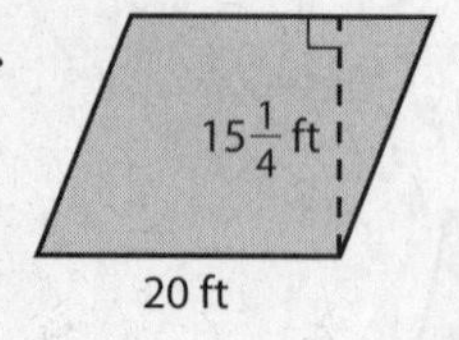

14. $3\frac{1}{2}$ yd

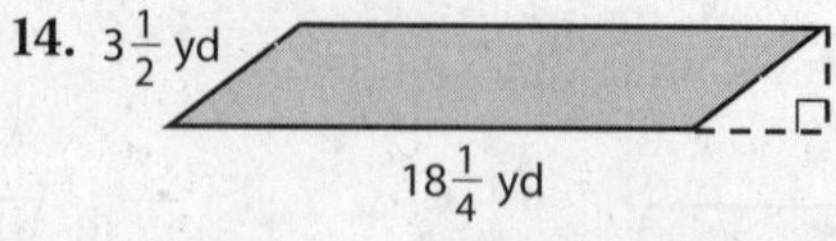

15.

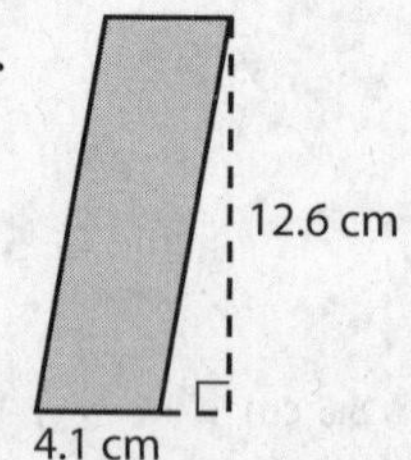

16. What is the measure of the area of a parallelogram with a base of $6\frac{2}{3}$ inches and a height of $1\frac{1}{2}$ inches?

17. Find the area of a parallelogram with base $7\frac{1}{5}$ yards and height $1\frac{1}{9}$ yards.

Math Skills Study Guide

Area of Triangles

The area A of a triangle is one half the product of any base b and its height h.

Symbols $A = \frac{1}{2}bh$

Model

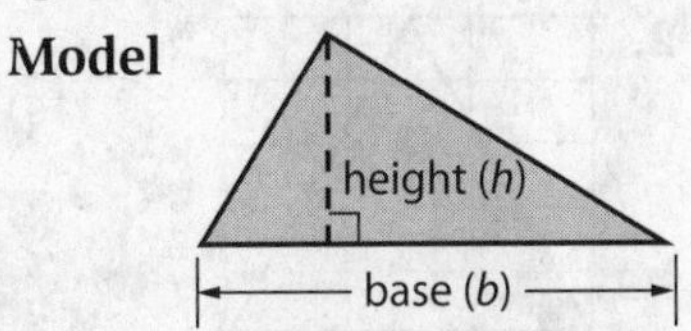

EXAMPLES **Find the area of each triangle.**

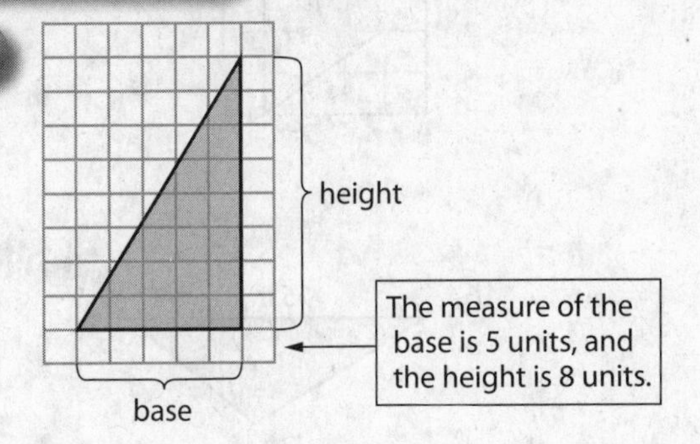

$A = \frac{1}{2}bh$ Area of a triangle

$A = \frac{1}{2}(5)(8)$ Replace b with 5 and h with 8.

$A = \frac{1}{2}(40)$ Multiply $5 \times 8 = 40$

$A = 20$

The area of the triangle is 20 square units.

2

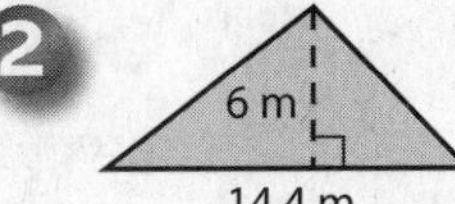

$A = \frac{1}{2}bh$

$A = \frac{1}{2}(14.4)(6)$ Replace b with 14.4 and h with 6

0.5 [×] 14.4 [×] 6 [ENTER] **43.2**

The area of the triangle is 43.2 square meters.

EXERCISES

Find the area of each triangle. Round decimals to the nearest tenth.

1.

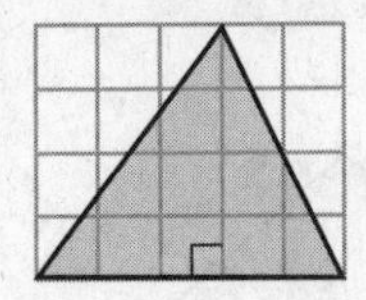

2.

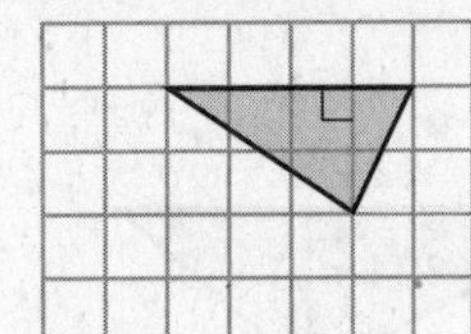

3.

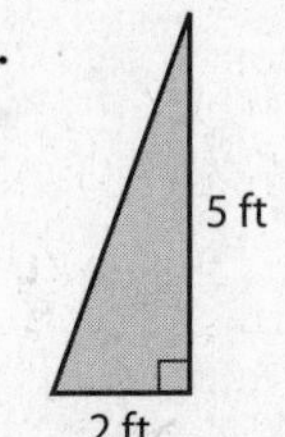

4.

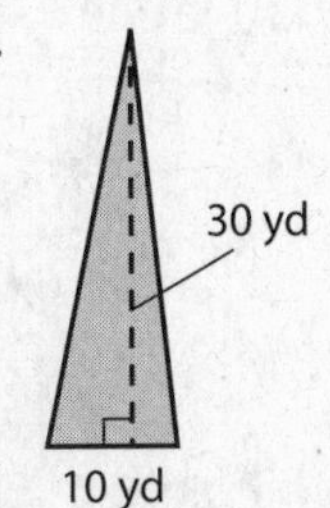

5.

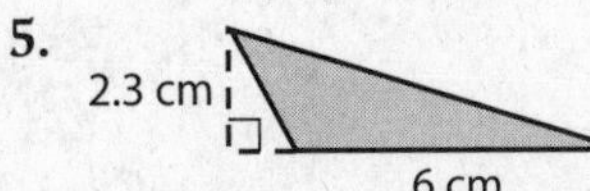

6.

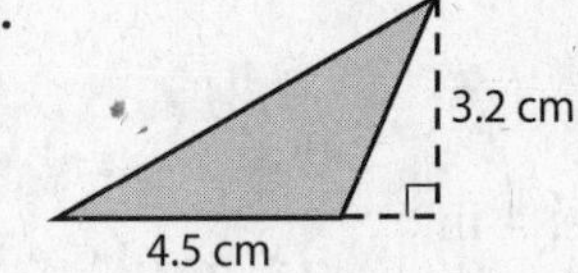

Name____________________ Date______ Period______

Math Skills Study Guide

Area of Triangles

Find the area of each triangle. Round decimals to the nearest tenth.

1. 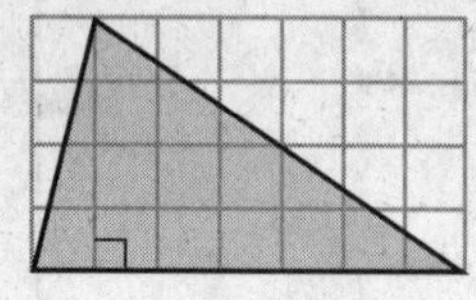

2. 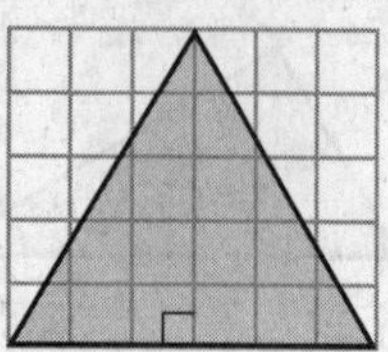

3. 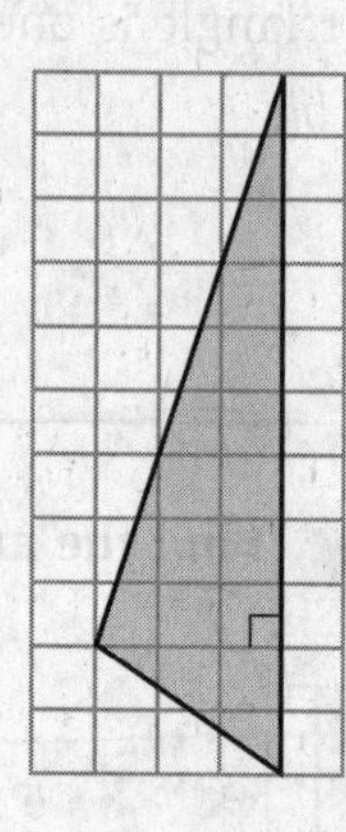

4. 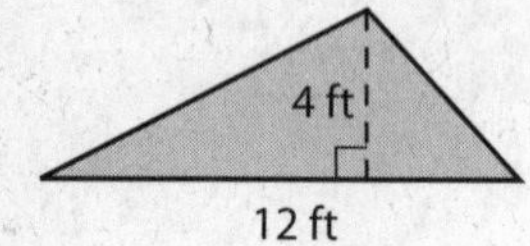

5. 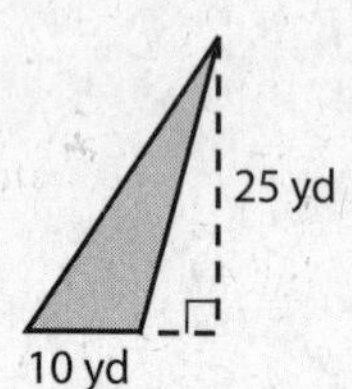

6. 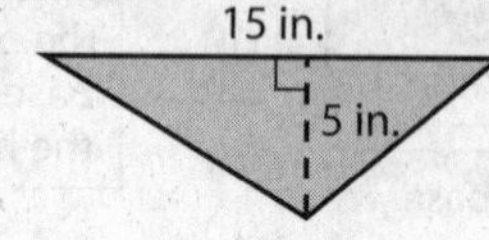

7. 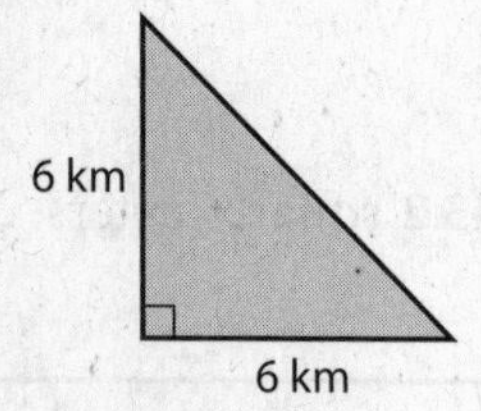

8. 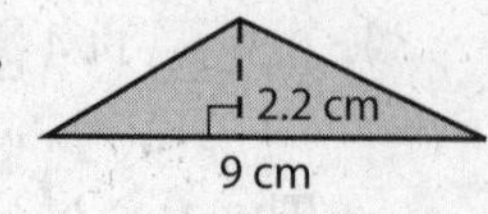

9. 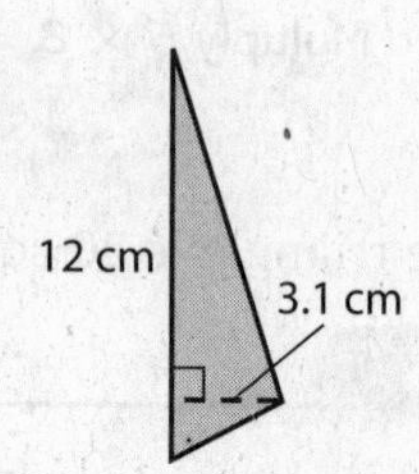

10. 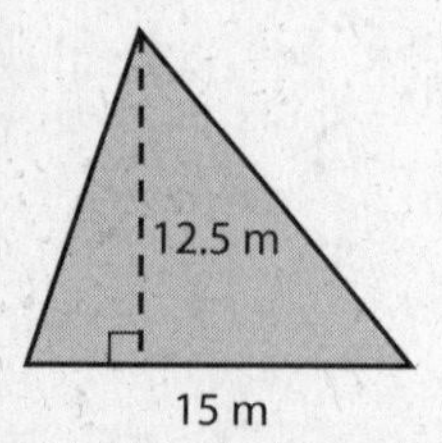

11. 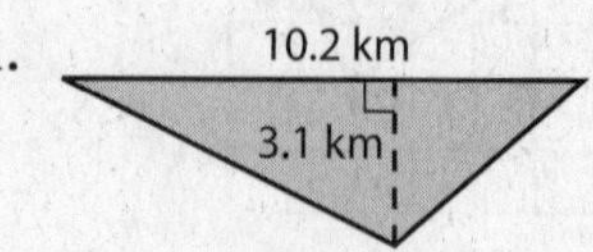

12. 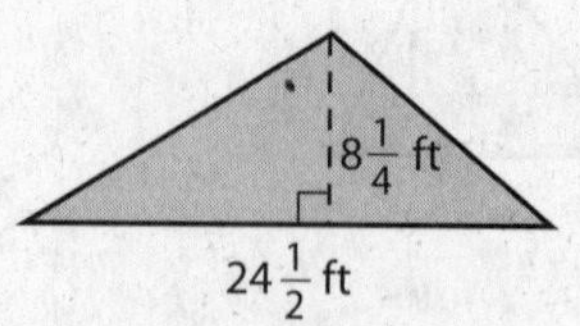

13. base: 4 in.
height: 11 in.

14. base: $4\frac{3}{4}$ yd
height: $1\frac{1}{3}$ yd

15. base: $5\frac{1}{4}$ ft
height: $2\frac{2}{3}$ ft

Name ______________________ Date ________ Period ________

Math Skills Study Guide

Area of Circles

The area A of a circle is the product of π and the square of the radius r.

Symbols $A = \pi r^2$

Model

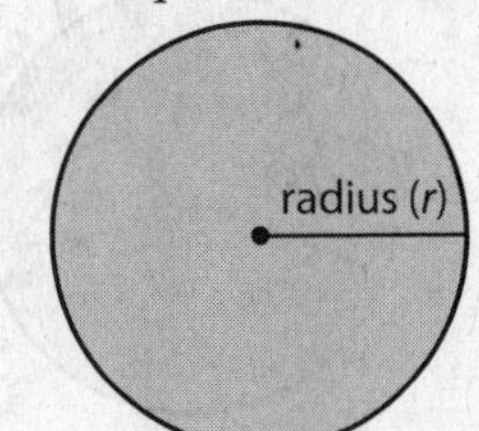

EXAMPLE 1 **Find the area of the circle to the nearest tenth.**

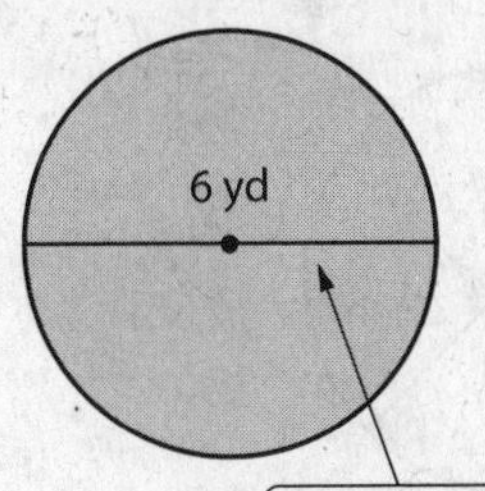

The diameter is 6 yards. So, the radius is 6 ÷ 2 or 3 yards.

$A = \pi r^2$	Area of a circle
$A \approx 3.14 \times 3^2$	Replace π with 3.14 and r with 3.
$A \approx 3.14 \times 9$	Evaluate 3^2.
$A \approx 28.26$	Use a calculator.

The area of the circle is about 28.3 square yards.

EXERCISES

Find the area of each circle to the nearest tenth. Use 3.14 for π.

1.

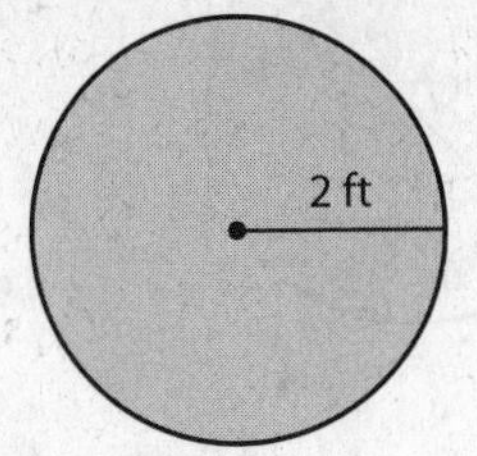

2.

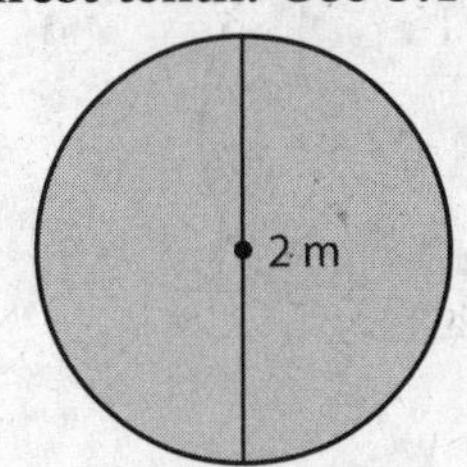

3.

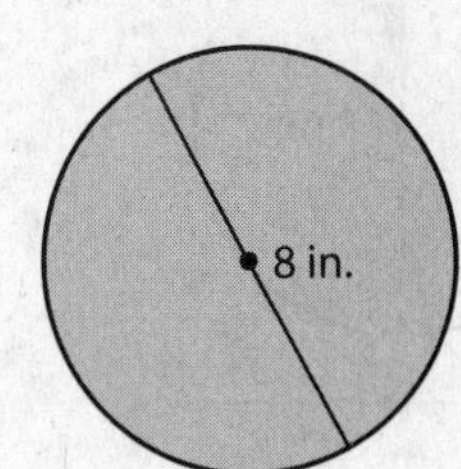

4.

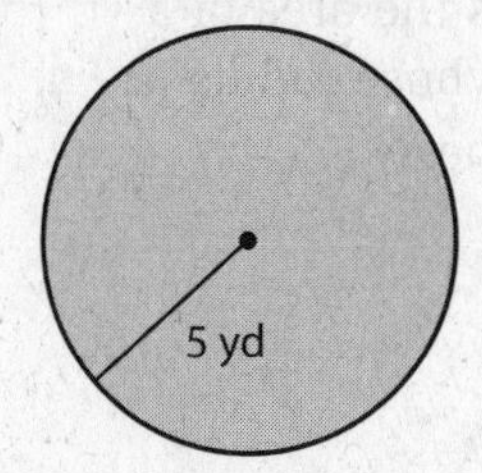

5.

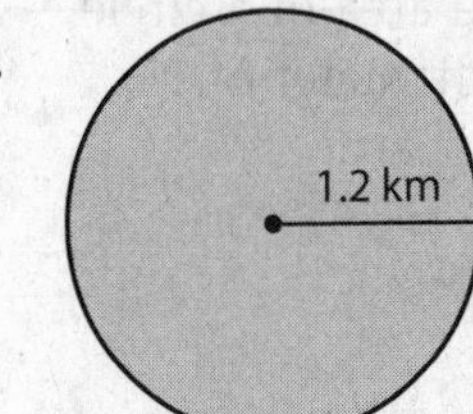

6.

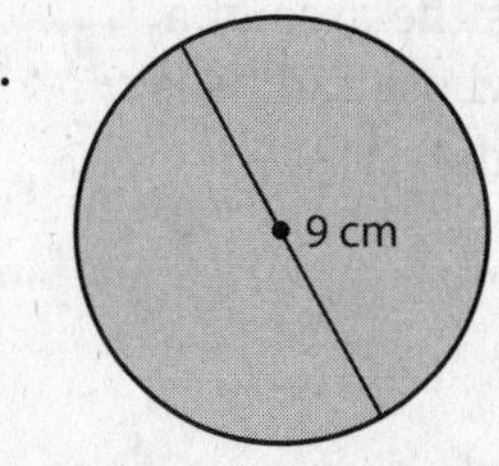

Name________________________ Date________ Period________

Math Skills Study Guide

Area of Circles

Find the area of each circle to the nearest tenth. Use 3.14 for π.

1.

2.

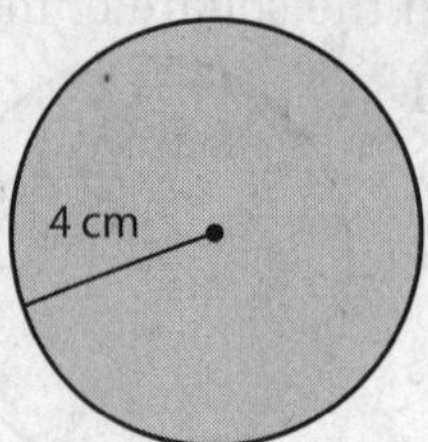

3.

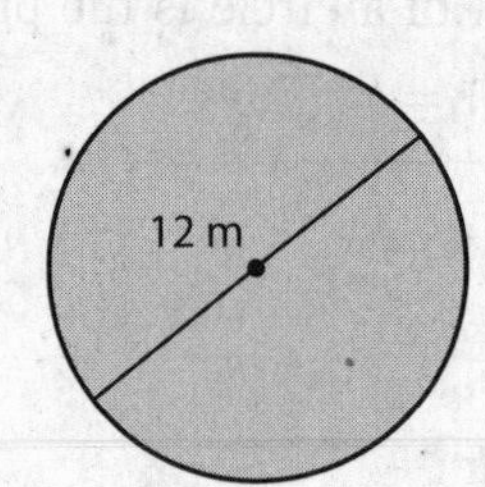

4.

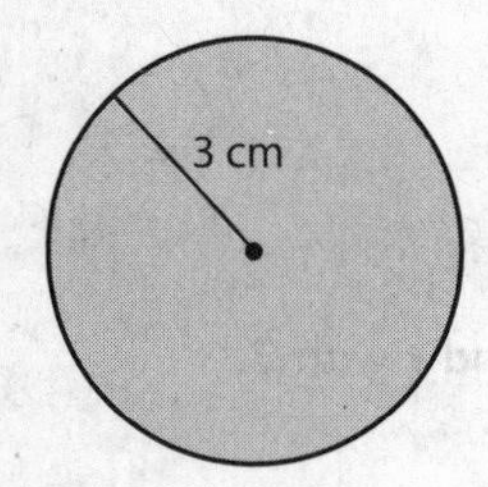

5.

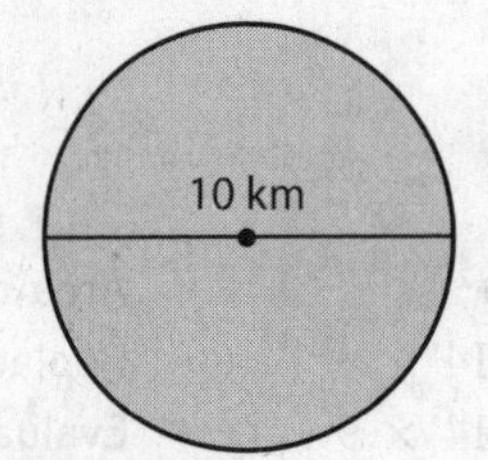

6.

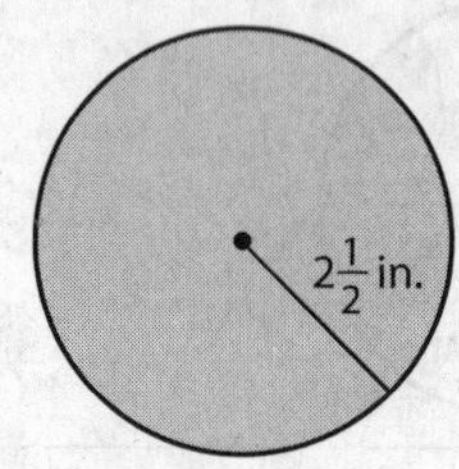

7.

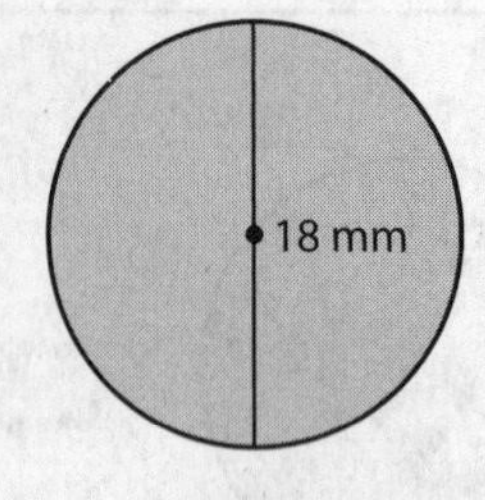

8.

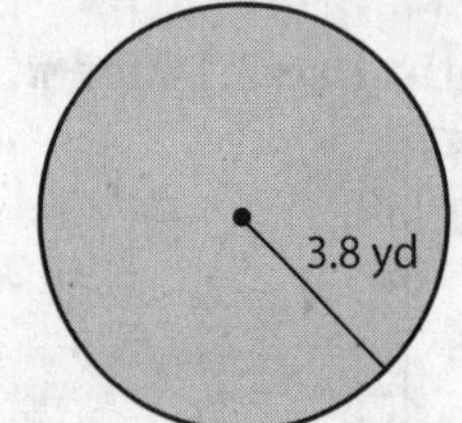

9.

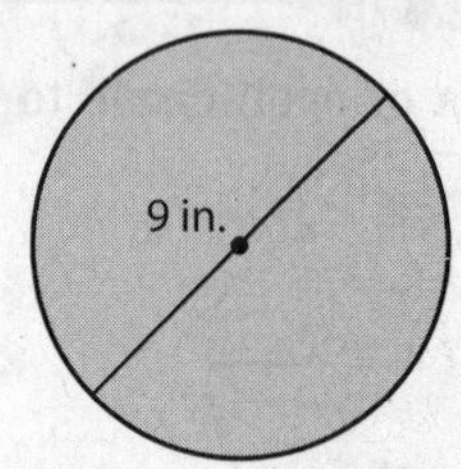

10. What is the area of a circle whose radius is 4.2 yards?

11. Find the area of a circle with a diameter of 13 meters.

12. What is the area of a circle whose radius is 6.6 inches?

Math Skills Study Guide

Surface Area of Rectangular Prisms

The surface area S of a rectangular prism with length ℓ, width w, and height h is the sum of the areas of the faces.

Symbols $S = 2\ell w + 2\ell h + 2wh$ **Model**

h

w

ℓ

EXAMPLE 1 **Find the surface area of the rectangular prism.**

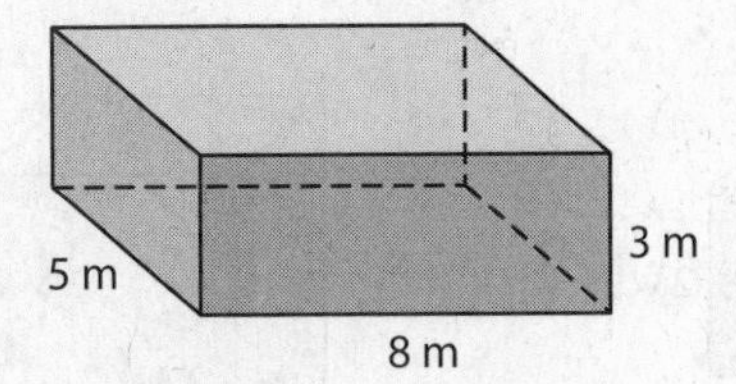

Find the area of each face.

top and bottom
$2(\ell w) = 2(8 \times 5) = 80$

front and back
$2(\ell h) = 2(8 \times 3) = 48$

two sides
$2(wh) = 2(5 \times 3) = 30$

Add to find the surface area. The surface area is $80 + 48 + 30$ or 158 square meters.

EXERCISES

Find the surface area of each rectangular prism. Round decimals to the nearest tenth.

1.

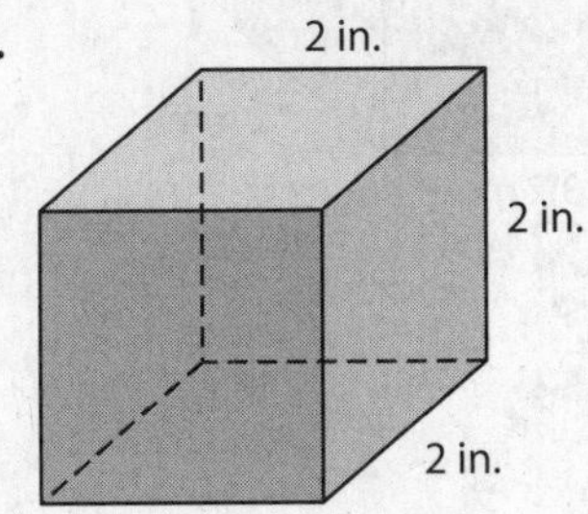

2.

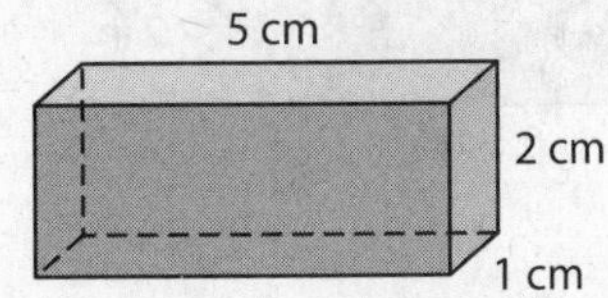

3.

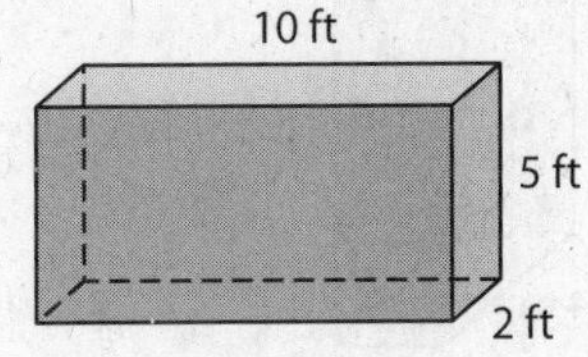

4.

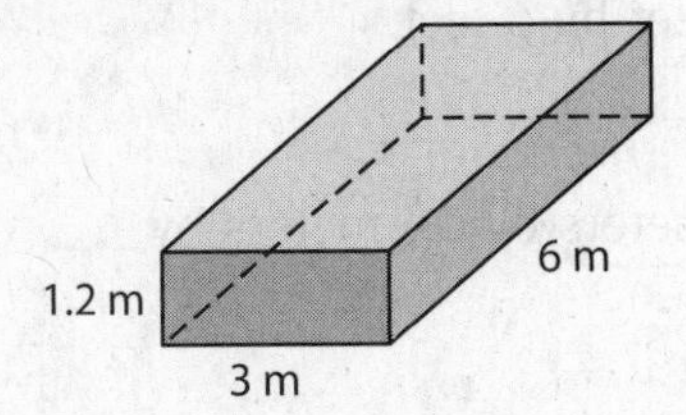

5.

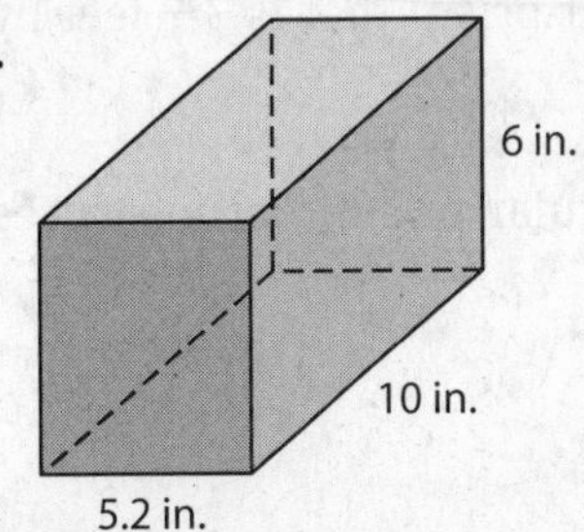

6.

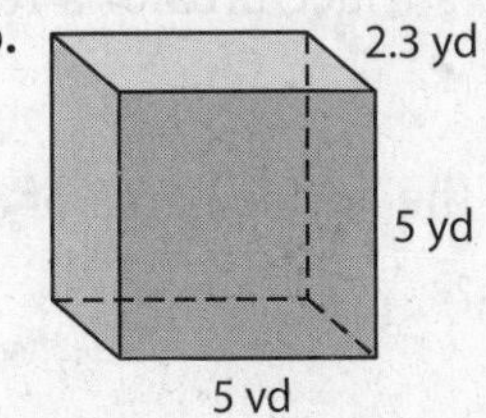

Name____________________________ Date________ Period________

Math Skills Study Guide

Surface Area of Rectangular Prisms

Find the surface area of each rectangular prism. Round decimals to the nearest tenth.

1.

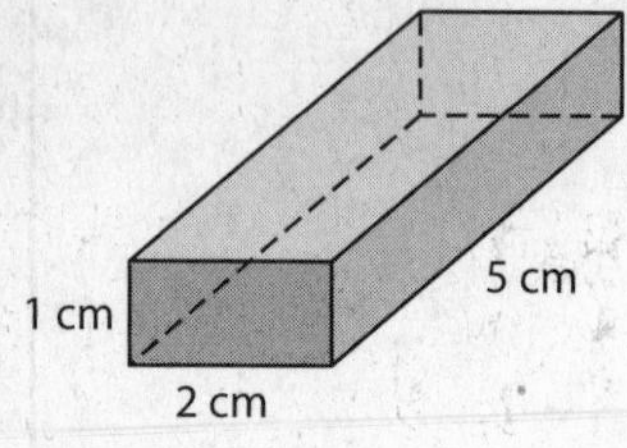

2.

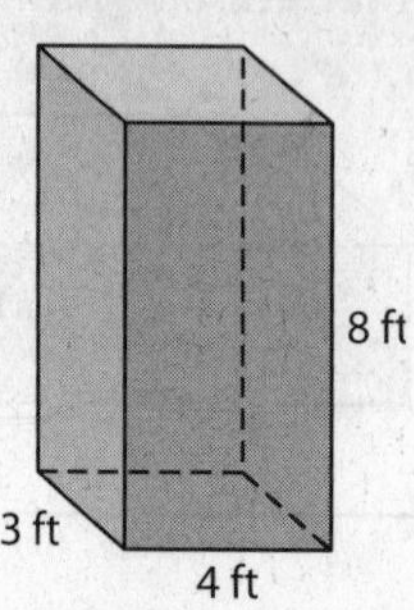

3.

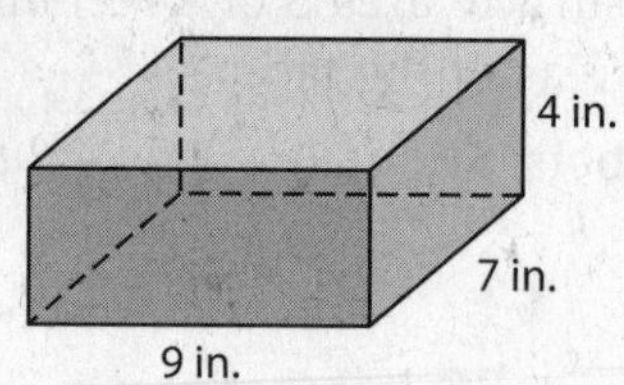

4.

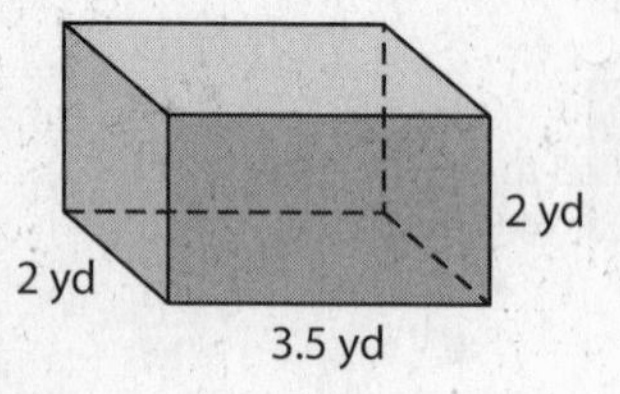

5.

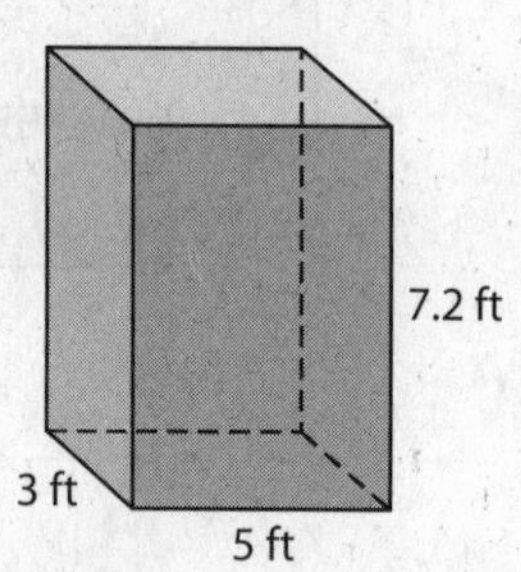

6.

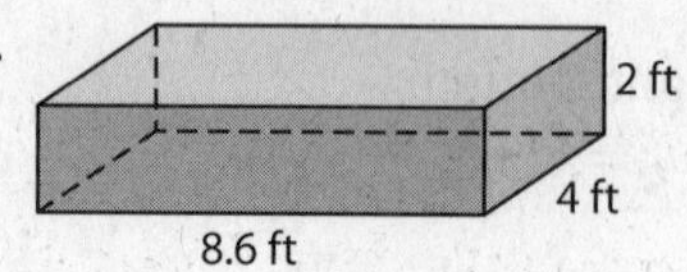

7.

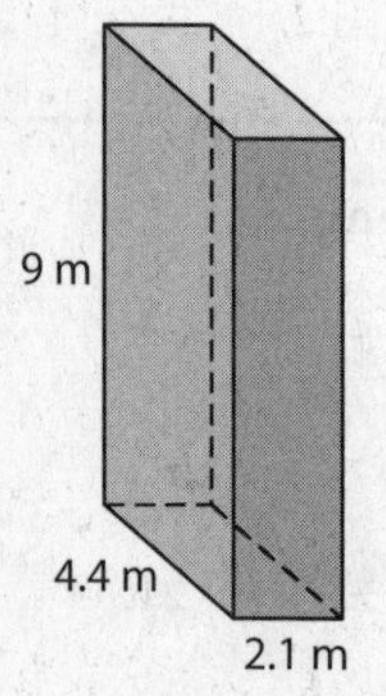

8.

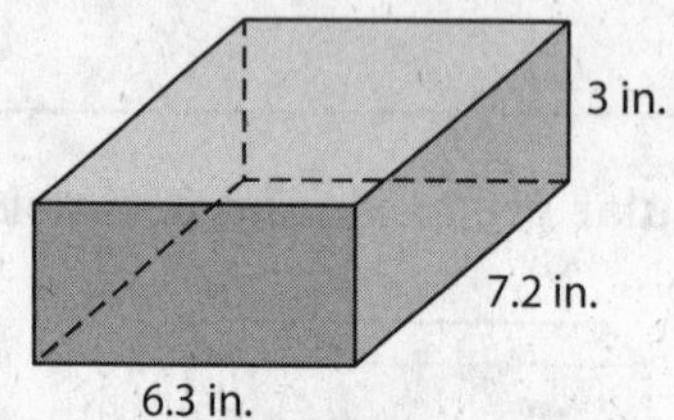

9.

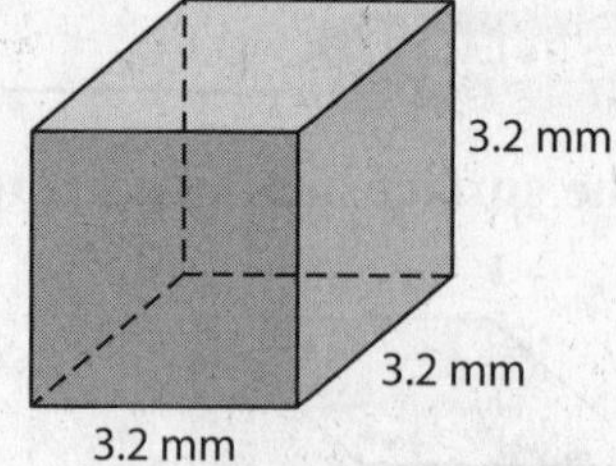

10. Find the surface area of a rectangular prism that is $3\frac{1}{2}$ feet by $4\frac{1}{4}$ feet by 6 feet.

11. What is the surface area of a rectangular prism that measures $2\frac{1}{3}$ meters by $2\frac{1}{2}$ meters by 4 meters?

Name ______________________ Date ________ Period ________

Math Skills Study Guide

Volume of Rectangular Prisms

The **volume** of a solid is the measure of space occupied by it. It is measured in cubic units such as cubic centimeters (cm^3) or cubic inches (in^3). The volume of the figure at the right can be shown using cubes.

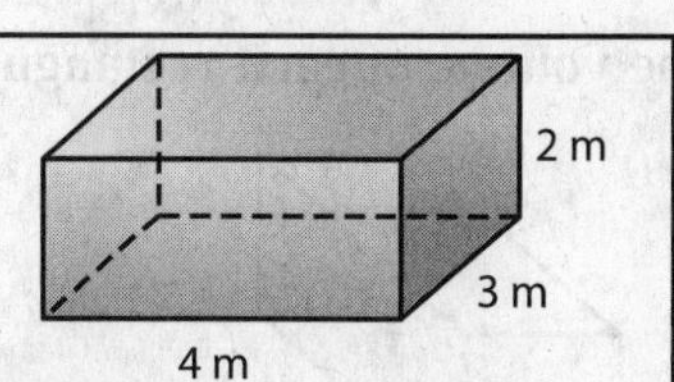

The bottom layer, or base, has $4 \cdot 3$ or 12 cubes. →

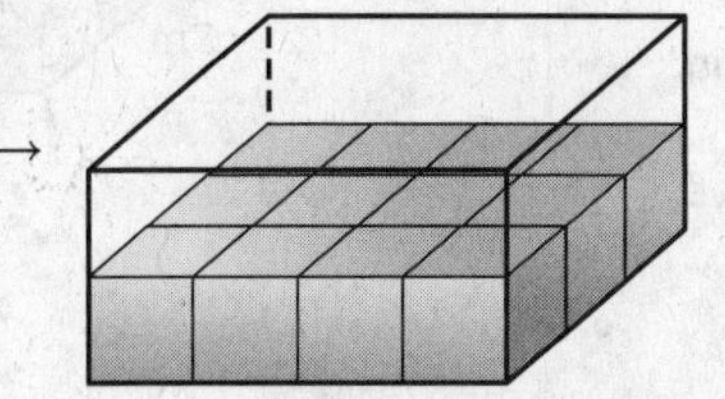

There are two layers

It takes 12 ? 2 or 24 cubes to fill the box. So, the volume of the box is 24 cubic meters.

A **rectangular prism** is a solid figure that has two parallel and congruent sides, or bases, that are rectangles. To find the volume of a rectangular prism, multiply the area of the base and the height, or find the product of the length ℓ, the width w, and the height h.

$V = Bh$ or $V = \ell wh$

EXAMPLE 1 **Find the volume of the rectangular prism.**

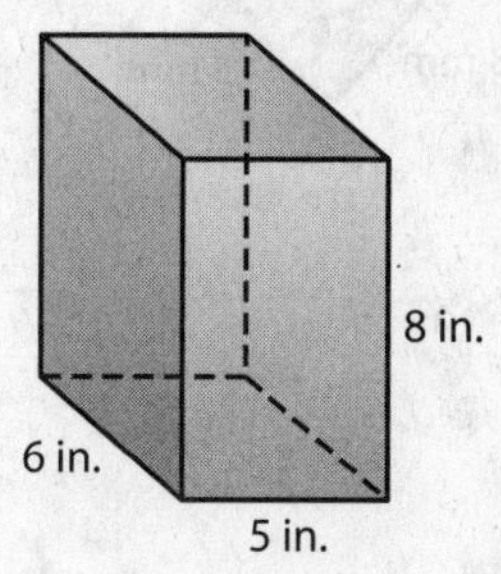

$V = \ell\text{wh}$	Volume of a rectangular prism
$V = 5 \cdot 6 \cdot 8$	Replace , with 5, *w* with 6, and *h* with 8.
$V = 240$	Multiply.

The volume is 240 cubic inches.

EXERCISES

Find the volume of each rectangular prism. Round decimals to the nearest tenth.

1.

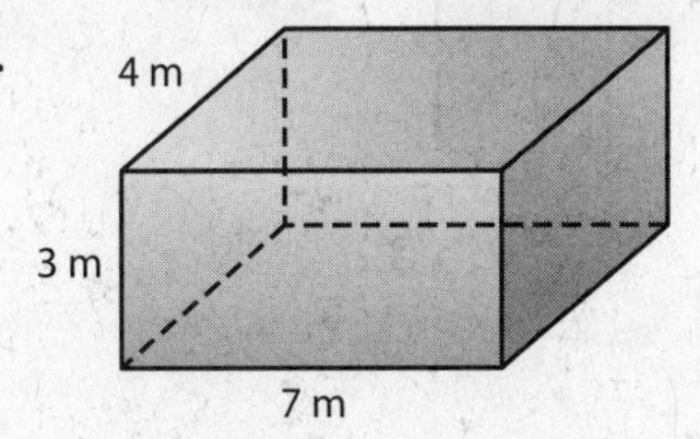

2.

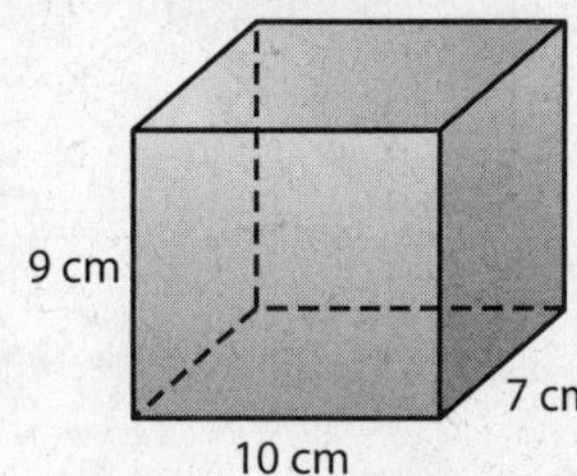

3.

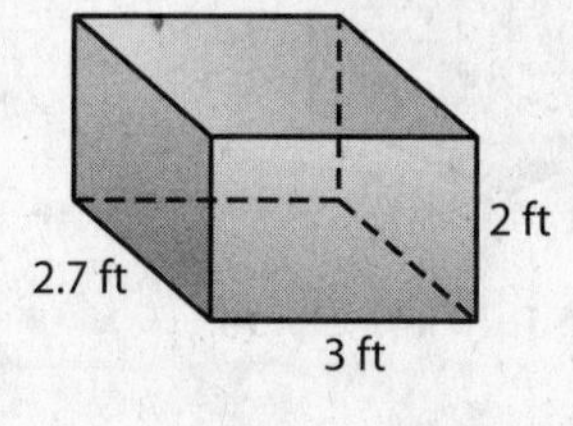

Name_________________________ Date _______ Period _______

Math Skills Study Guide

Volume of Rectangular Prisms

Find the volume of each rectangular prism. Round decimals to the nearest tenth.

1.

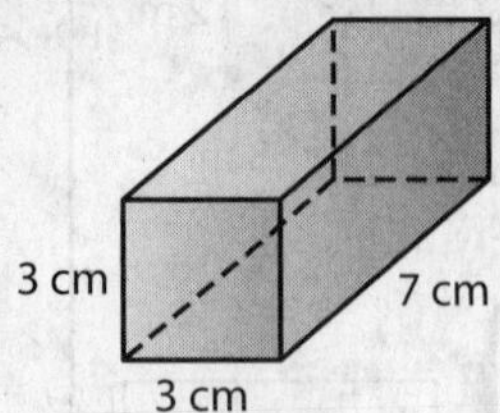

2.

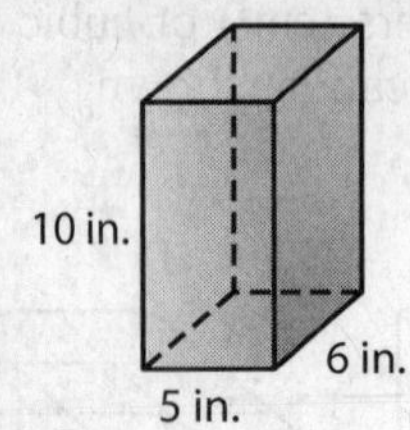

3.

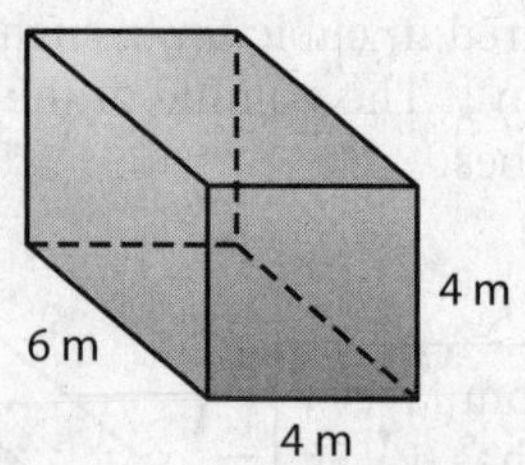

4.

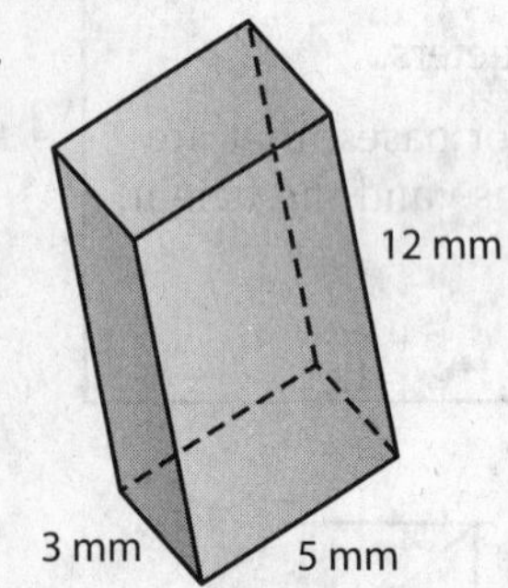

5.

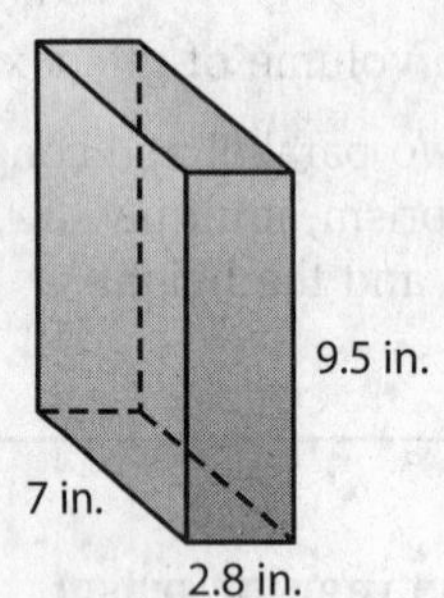

6.

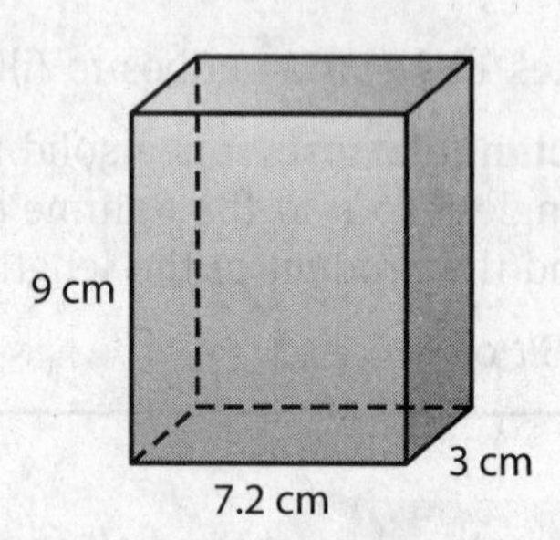

7.

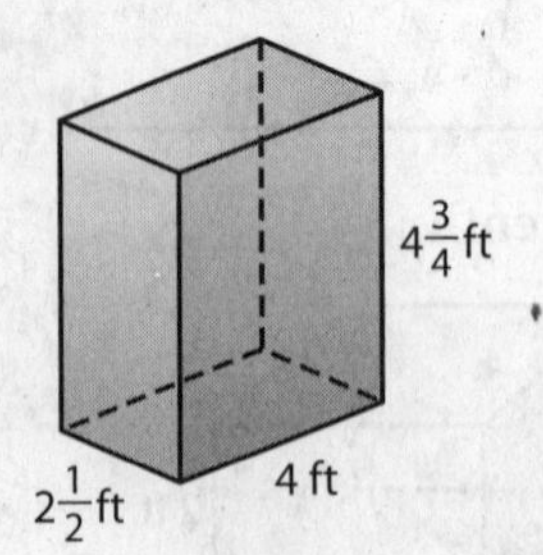

8.

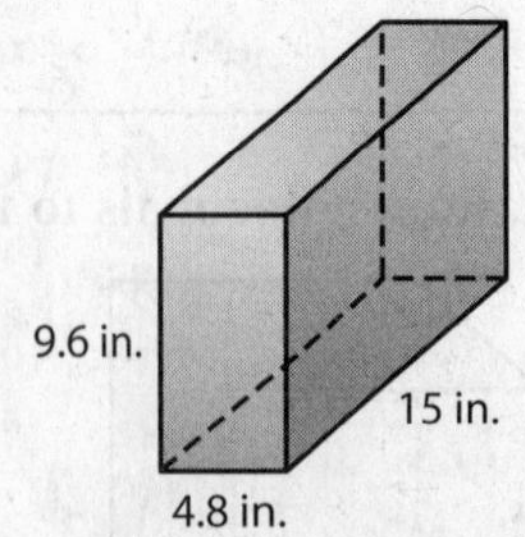

9.

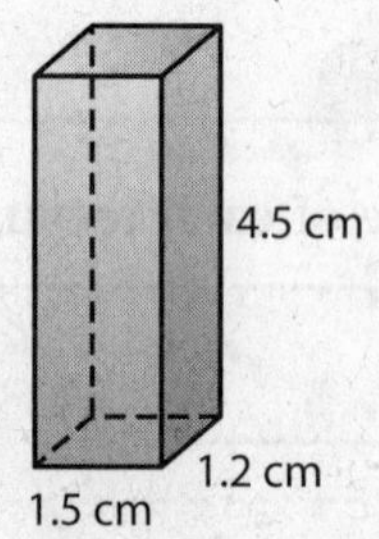

Name ______________________ Date ________ Period ________

Math Skills Study Guide

Using a Measurement Conversion Chart

You may sometimes want to convert customary measurements to metric measurements. For example, suppose you are reading about horses and want to know how long 5 furlongs are.

Start by finding a conversion table such as the one shown here. (Dictionaries often include such tables.)

1 mil = 0.001 inch	= 0.0254 millimeter
1 inch = 1,000 mil	= 2.54 centimeters
12 inches = 1 foot	= 0.3048 meter
3 feet = 1 yard	= 0.9144 meter
$5\frac{1}{2}$ yards, or $16\frac{1}{2}$ feet = 1 rod	= 5.029 meters
40 rods = 1 furlong	= 201.168 meters
8 furlongs, 5,280 feet, 1,760 yards = 1 (statute) mile	= 1.6093 kilometers
3 miles = 1 (land) league	= 4.828 kilometers

To change from a large unit to a small unit, multiply. To change from a small unit to a large one, divide.

EXAMPLE 1 **Change 5 furlongs to meters.**

1 furlong = 201.168 meters
5 furlongs = 1,005.84 meters
So, 5 furlongs is about 1,000 meters, or 1 kilometer.

Change each measurement to a metric measurement. Round each answer to the nearest tenth.

1. 10 yards

2. 100 leagues

3. 10 inches

4. 100 rods

5. 1,000 mils

6. 10 feet

7. 50 miles

8. 50 furlongs

9. 50 inches

10. 200 feet

11. 200 miles

12. 200 yards

Name________________________ Date________ Period________

Math Skills Study Guide

Scale Drawings

A **scale drawing** represents something that is too large or too small to be drawn or built at actual size. Similarly, a **scale model** can be used to represent something that is too large or built too small for an actual-size model. The **scale** gives the relationship between the drawing/model measure and the actual measure.

EXAMPLE **On this map, each grid unit represents 50 yards. Find the distance from Patrick's Point to Agate Beach.**

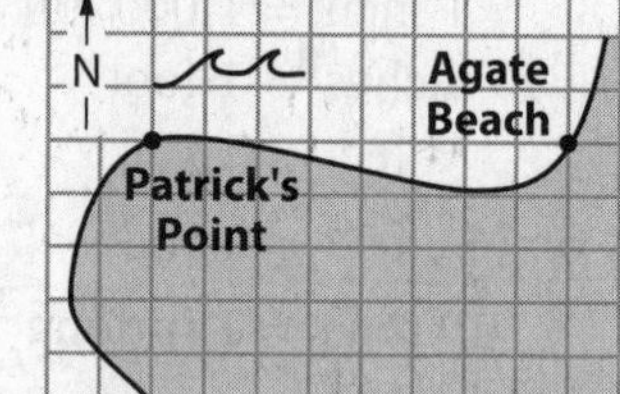

1 unit = 50 yards ⟵ Scale
× 8   × 8 ⟵ Multiply both sides by 8.
8 units = 400 yards

It is 400 yards from Patrick's Point to Agate Beach.

EXERCISES

Find the actual distance between each pair of cities. Round to the nearest tenth if necessary.

	Cities	Map Distance	Scale	Actual Distance
1.	Los Angeles and San Diego, California	6.35 cm	1 cm = 20 mi	
2.	Lexington and Louisville, Kentucky	15.6 cm	1 cm = 5 mi	
3.	Des Moines and Cedar Rapids, Iowa	16.27 cm	2 cm = 15 mi	
4.	Miami and Jacksonville, Florida	11.73 cm	$\frac{1}{2}$ cm = 20 mi	

Suppose you are making a scale drawing. Find the length of each object on the scale drawing with the given scale. Then find the scale factor.

5. an automobile 16 feet long; 1 inch:6 inches

6. a lake 85 feet across; 1 inch = 4 feet

7. a parking lot 200 meters wide; 1 centimeter:25 meters

8. a flag 5 feet wide; 2 inches = 1 foot

Name ______________________ Date ________ Period ________

Math Skills Study Guide

Scale Drawings

ARCHITECTURE The scale on a set of architectural drawings for a house is $\frac{1}{2}$ inch = $1\frac{1}{2}$ feet. Find the length of each part of the house.

	Room	Drawing Length	Actual Length
1.	Living Room	5 inches	
2.	Dining Room	4 inches	
3.	Kitchen	$5\frac{1}{2}$ inches	
4.	Laundry Room	$3\frac{1}{4}$ inches	
5.	Basement	10 inches	
6.	Garage	$8\frac{1}{3}$ inches	

ARCHITECTURE As part of a city building refurbishment project, architects have constructed a scale model of several city buildings to present to the city commission for approval. The scale of the model is 1 inch = 9 feet.

7. The courthouse is the tallest building in the city. If it is $7\frac{1}{2}$ inches tall in the model, how tall is the actual building?

8. The city commission would like to install new flagpoles that are each 45 feet tall. How tall are the flagpoles in the model?

9. In the model, two of the flagpoles are 4 inches apart. How far apart will they be when they are installed?

10. The model includes a new park in the center of the city. If the dimensions of the park in the model are 9 inches by 17 inches, what are the actual dimensions of the park?

11. Find the scale factor.

Name ______________________ Date ________ Period ________

Math Skills Study Guide

Measures of Center

When working with numerical data, it is often helpful to use one or more numbers to represent the whole set. These numbers are called the measures of center. You will study the mean, median, and mode.

Statistic	Definition
mean	sum of the data divided by the number of values in the data set
median	middle number of the ordered data, or, if there are an even number of values, the mean of the middle two numbers
mode	number or numbers that occur most often

EXAMPLE **Jason recorded the number of hours he spent watching television each day for a week. Find the mean, median, and mode for the number of hours.**

Mon.	Tues.	Wed.	Thurs.	Fri.	Sat.	Sun.
2	3.5	3	0	2.5	6	4

$$\text{mean} = \frac{\text{sum of hours}}{\text{number of days}}$$

$$= \frac{2 + 3.5 + 3 + \ldots + 4}{7}$$

$= 3$ The mean is 3 hours.

To find the median, order the numbers from least to greatest and locate the number in the middle.

0 2 2.5 (3) 3.5 4 6 The median is 3 hours.

There is no mode because each number occurs once in the set.

EXERCISES

Find the mean, median, and mode for each set of data.

1. Maria's test scores
92, 86, 90, 74, 95, 100, 90, 50

2. Rainfall last week in inches
0, 0.3, 0, 0.1, 0, 0.5, 0.2

3. Resting heart rates of 8 males
84, 59, 72, 63, 75, 68, 72, 63

Name ____________________ Date ______ Period ______

Math Skills Study Guide

Measures of Center

Find the mean, median, and mode for each set of data. Round decimals to the nearest tenth.

1. 6, 3, 3, 12, 13, 15, 7

2. 1, 1, 0, 2, 1, 1, 0, 0, 1

3. 202, 195, 219, 220

4. 2.5, 4.0, 8.7, 3.3, 3.3, 5.2

5. 21, 23, 39, 44, 27, 25, 28, 30

6. 87, 85, 87, 87, 87

Find the mean, median, and mode for each set of data. Round decimals to the nearest tenth.

7.

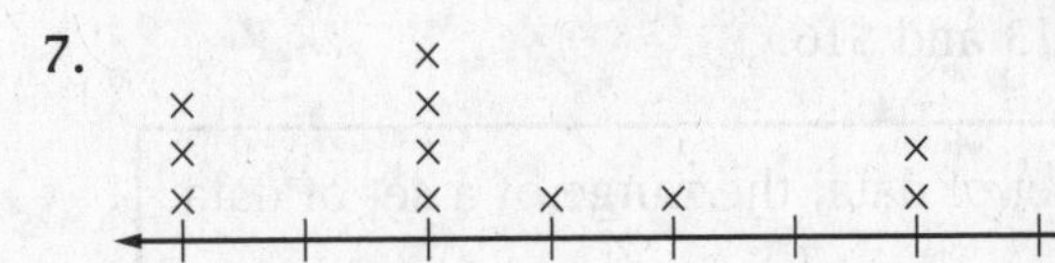

8.

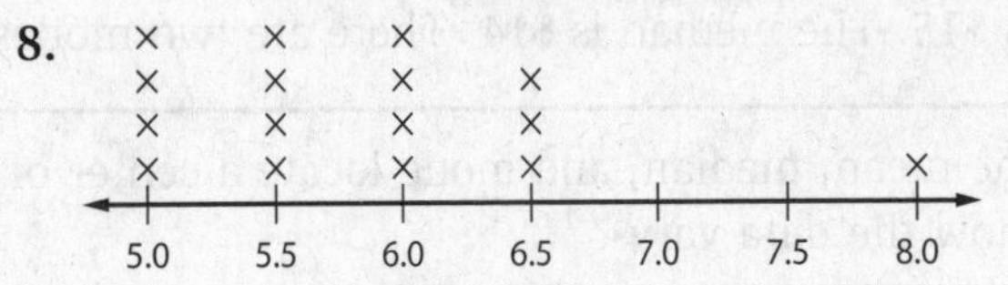

9. TEMPERATURE The average daily temperature by month for one year in Denver, Colorado, is given below. Find the mean, median, and mode for temperature.

Month	Jan	Feb	Mar	Apr	May	June	July	Aug	Sept	Oct	Nov	Dec
Temp. (°F)	43°	47°	51°	61°	71°	82°	88°	86°	78°	67°	52°	46°

Source: *The Universal Almanac*

Name ______________________ Date ________ Period ________

Math Skills Study Guide

Median, Mode, and Range

The **median** is the middle number of the data put in order, or, if there are an even number of values, the mean of the middle two numbers. The **mode** is the number or numbers that occur most often.

EXAMPLE 1 **The table shows the costs of seven different books. Find the mean, median, and mode of the data.**

Book Costs ($)			
22	13	11	16
14	13	16	

mean: $\frac{22 + 13 + 11 + 16 + 14 + 13 + 16}{7} = \frac{105}{7}$ or 15

To find the median, write the data in order from least to greatest.
median: 11, 13, 13, (14), 16, 16, 22

To find the mode, find the number or numbers that occur most often.
mode: 11, (13, 13), 14, (16, 16), 22

The mean is $15. The median is $14. There are two modes, $13 and $16.

Whereas the mean, median, and mode locate a center of a set of data, the **range** of a set of data describes how the data vary.

EXAMPLE 2 **Find the range of the data in the table. Then write a sentence describing how the data vary.**

Temperature (°F)		
40	32	55
60	63	50

The greatest value is 63. The smallest value is 32. So, the range is 63° − 32°or 31°.

EXERCISES

Find the mean, median, mode, and range of each set of data.

1. 14, 13, 14, 16, 8

2. 29, 31, 14, 21, 31, 22, 20

3.

Quiz Scores		
72	60	80
68	72	86

4.

Snowfall (in.)			
2	6	5	4
3	0	1	

Name______________________________ Date________ Period________

Math Skills Study Guide

Measures of Center

Find the mean, median, and mode for each set of data. Round decimals to the nearest tenth.

1. 4, 6, 12, 5, 8

2. 16, 18, 15, 16, 21, 16

3. 55, 46, 50, 42, 39

4. 17, 16, 13, 17, 17, 10, 10, 13, 10

5. 25, 25, 25, 20

6. 3.1, 4.5, 4.5, 4.3, 6.0, 3.2

Find the mean, median, and mode for each set of data. Round decimals to the nearest tenth.

7.

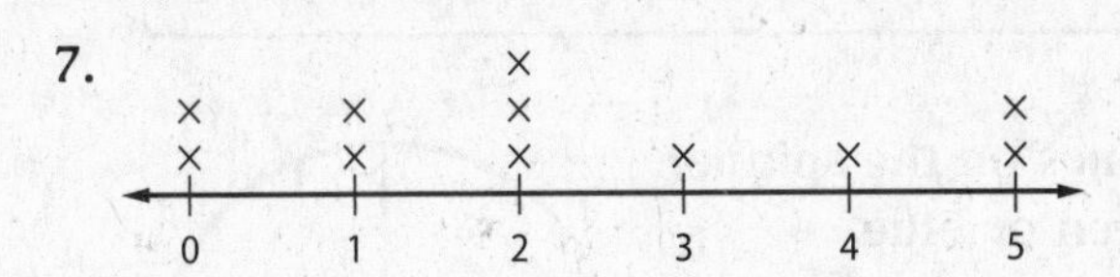

8.

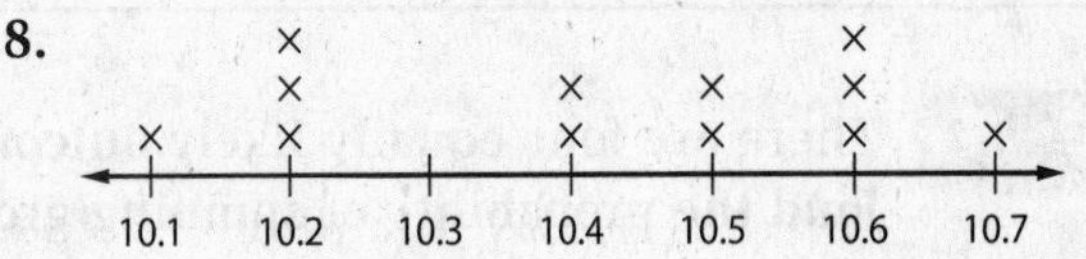

9. TORNADOES The table below shows the number of tornadoes reported in the United States from 1980–1990. Find the mean, median, and mode for the number of tornadoes.

Year	1980	1981	1982	1983	1984	1985	1986	1987	1988	1989	1990
Number of Tornadoes	866	783	1046	931	907	684	764	656	702	858	1132

Source: *The Universal Almanac*

Name ______________________ Date ________ Period ________

Math Skills Study Guide

Theoretical Probability

When tossing a coin, there are two possible **outcomes,** heads and tails. Suppose you are looking for heads. If the coin lands heads-up, this would be a favorable outcome or **event.** When outcomes are equally likely, you can use a ratio to find probability. The probability of an event is a number from 0 to 1, including 0 and 1. The closer a probability is to 1, the more likely it is to happen.

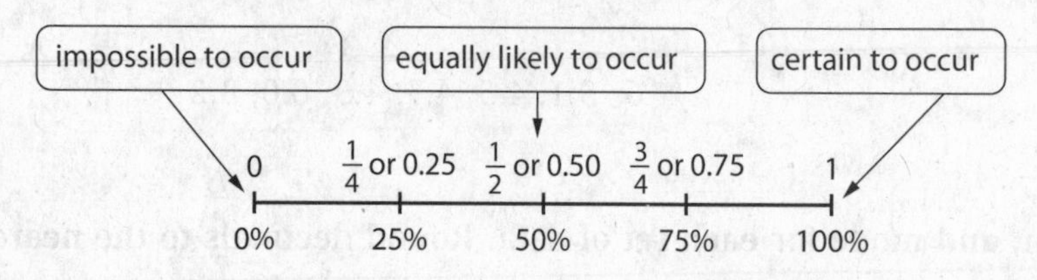

EXAMPLE 1 **There are four equally likely outcomes on the spinner. Find the probability of spinning green or blue.**

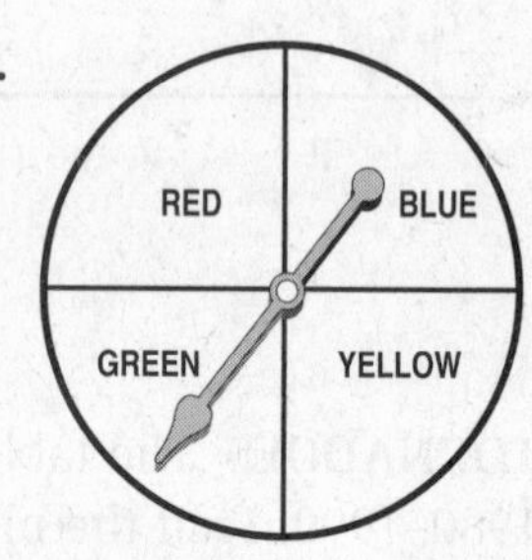

$$P(\text{green or blue}) = \frac{\text{number of favorable outcomes}}{\text{number of possible outcomes}}$$

$$= \frac{2}{4} \text{ or } \frac{1}{2}$$

The probability of landing on green or blue is $\frac{1}{2}$, 0.50, or 50%.

Complementary events are two events in which either one or the other must happen, but both cannot happen. The sum of the probabilities of complementary events is 1.

EXAMPLE 2 **There is a 25% chance that Sam will win a prize. What is the probability that Sam will not win a prize?**

$P(\text{win}) + P(\text{not win}) = 1$

$0.25 + P(\text{not win}) = 1$ — Replace P(win) with 0.25.

$-0.25 \qquad = -0.25$ — Subtract 0.25 from each side.

$P(\text{not win}) = 0.75$

So, the probability that Sam won't win a prize is 0.75, 75%, or $\frac{3}{4}$.

EXERCISES

1. There is a 90% chance that it will rain. What is the probability that it will not rain?

One pen is chosen without looking from a bag that has 3 blue pens, 6 red, and 3 green. Find the probability of each event. Write each answer as a fraction, a decimal, and a percent.

2. P(green)

3. P(blue or red)

4. P(yellow)

Name________________________________ Date________ Period________

Math Skills Study Guide

Theoretical Probability

One of these cards is randomly chosen. Find each probability. Write each answer as a fraction, a decimal, and a percent.

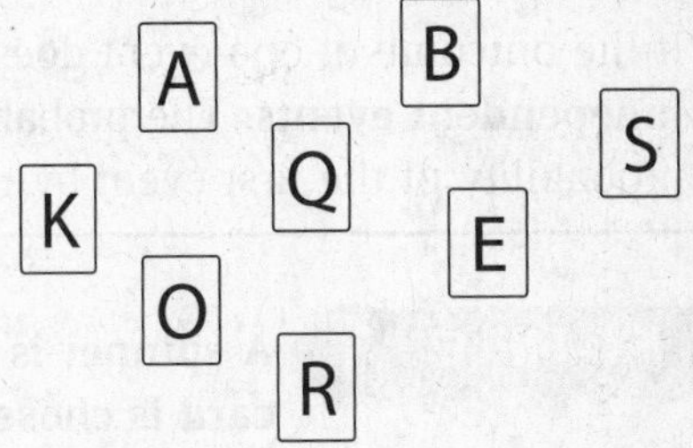

1. $P(\text{B})$

2. $P(\text{Q or R})$

3. $P(\text{vowel})$

4. $P(\text{consonant or vowel})$

5. $P(\text{consonant or A})$

6. $P(\text{T})$

The spinner shown is spun once. Find each probability

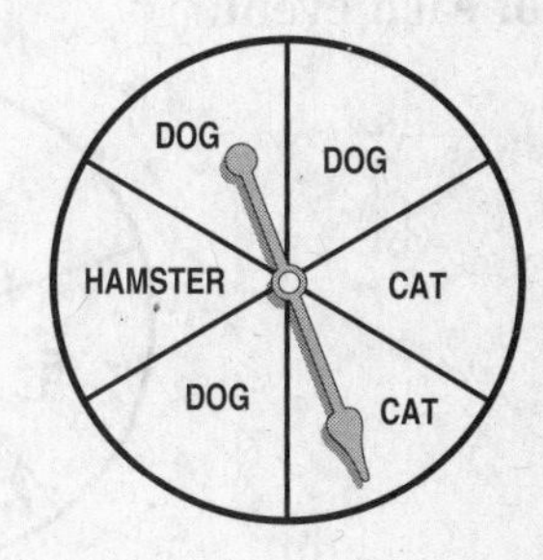

7. P(dog)

8. $P(\text{hamster})$

9. $P(\text{dog or cat})$

10. $P(\text{bird})$

11. $P(\text{mammal})$

WEATHER The weather reporter says that there is a 12% chance that it will be moderately windy tomorrow.

12. What is the probability that it will not be moderately windy?

Name ________________________ Date ______ Period ______

Math Skills Study Guide

Probability and Independent Events

If the outcome of one event does not affect the outcome of a second event, the two events are **independent events.** The probability of two independent events is found by multiplying the probability of the first event by the probability of the second event.

EXAMPLE 1 **A spinner is spun and a number card is chosen at random. What is the probability that red is spun and a 4 is chosen?**

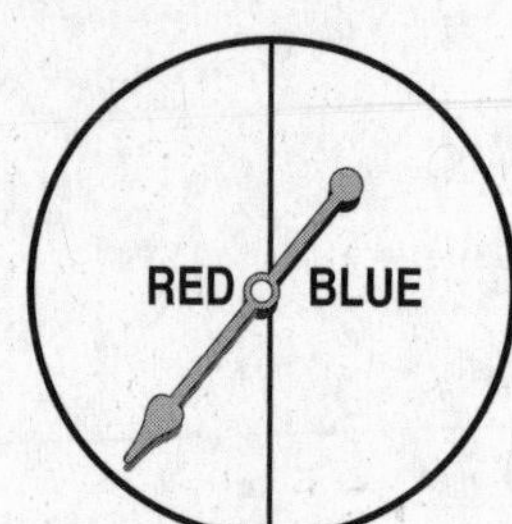

4	2	1
3	4	4

$P(\text{red}) = \frac{1}{2}$ $\quad P(4) = \frac{3}{6}$ or $\frac{1}{2}$

$P(\text{red and } 4) = \frac{1}{2} \times \frac{1}{2} = \frac{1}{4}$

So, the probability is $\frac{1}{4}$, 0.25, or 25%.

EXERCISES

A spinner is spun and a number card is chosen at random. Find the probability of each event.

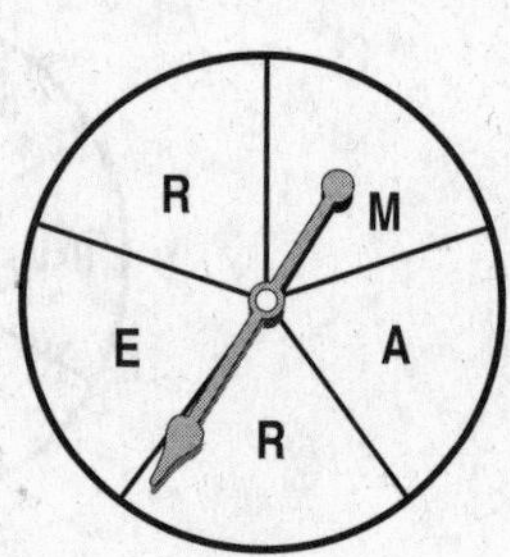

1	3	4
3	6	3

1. P(M and 3) **2.** P(R and 3) **3.** P(consonant and odd)

4. P(consonant and 3) **5.** P(vowel and less than 7) **6.** P(vowel and even)

A coin is tossed and a six-sided die is rolled. Find the probability of each event.

7. P(tails and even number) **8.** P(heads and number less than 4)

9. P(heads and number greater than 2)

Name________________________________ Date________ Period________

Math Skills Study Guide

Probability and Independent Events

The two spinners shown are spun. Find the probability of each event.

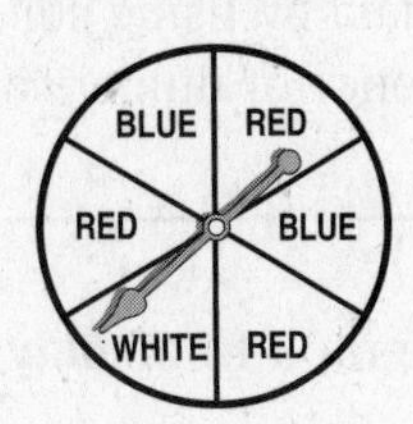

1. P(1 and white)

2. P(3 and red)

3. P(2 and blue)

4. P(odd and red)

5. P(4 and white)

6. P(even and any color other than white)

Suppose you select a card from the pile shown and then roll a number cube. Find the probability of each event.

P B B E T

7. P(B and 4)

8. P(B and even)

9. P(consonant and 5)

10. P(vowel and odd)

11. P(E and number less than 7)

12. P(5 and odd)

13. **LEAVES** The table lists the autumn leaves each girl collected. Each girl reaches into her own bag and randomly selects a leaf. Find the probability that Jane chooses a maple and Mary chooses an aspen leaf.

Name	Maple	Cottonwood	Aspen
Jane	14	8	6
Marry	8	10	2

Name ______________________ Date ________ Period ________

Math Skills Study Guide

Bar Graphs and Histograms

A **bar graph** is one method of comparing data by using solid bars to represent quantities. A **histogram** uses bars to represent the frequency of numerical data that have been organized into intervals.

EXAMPLE 1 **SIBLINGS Make a bar graph to display the data in the table below.**

Student	Number of Siblings
Isfu	1
Sue	6
Margarita	3
Akira	2

Step 1 Draw a horizontal and a vertical axis. Label the axes as shown. Add a title.

Step 2 Draw a bar to represent each student. In this case, a bar is used to represent the number of siblings for each student.

EXAMPLE 2 **SIBLINGS The number of siblings of 17 students have been organized into a table. Make a histogram of the data.**

Number of Siblings	Frequency
0–1	4
2–3	10
4–5	2
6–7	1

Step 1 Draw and label horizontal and vertical axes. Add a title.

Step 2 Draw a bar to represent the frequency of each interval.

EXERCISES

1. Make a bar graph for the data in the table.

Student	Number of Free Throws
Luis	6
Laura	10
Opal	4
Gad	14

2. Make a histogram for the data in the table.

Number of Free Throws	Frequency
0–1	1
2–3	5
4–5	10
6–7	4

Name_________________________________ Date_______ Period_______

Math Skills Study Guide

Bar Graphs and Line Graphs

Make a bar graph for each set of data.

1.

Cars Made in 2000	
Country	**Cars (millions)**
Brazil	1
Japan	8
Germany	5
Spain	2
U.S.A.	6

2.

People in America in 1630	
Colony	**People (hundreds)**
Maine	4
New Hampshire	5
Massachusetts	9
New York	4
Virginia	25

Use the bar graph made in Exercise 1.

3. Which country made the greatest number of cars?

4. How does the number of cars made in Japan compare to the number made in Spain?

For Exercises 5 and 6, make a line graph for each set of data.

5.

Yuba Country, California	
Year	**Population (thousands)**
1990	59
1992	61
1994	62
1996	61
1998	60
2000	60

6.

Everglades National Park	
Month	**Rainfall (inches)**
January	2
February	2
March	2
April	2
May	7
June	10

7. **POPULATION** Refer to the graph made in Exercise 5. Describe the change in Yuba Country's population from 1990 to 2000.

8. **WEATHER** Refer to the graph made in Exercise 6. Describe the change in the amount of rainfall from January to June.

Name ____________________ Date ________ Period ________

Math Skills Study Guide

Bar Graphs and Histograms

ZOOS For Exercises 1 and 2, use the table. It shows the number of species at several zoological parks.

Zoo	Species
Los angeles	350
Lincoln Park	290
Cincinnati	700
Bronx	530
Oklahoma City	600

1. Make a bar graph of the data.

Animal Species in Zoos

2. Which zoological park has the most species?

ZOOS For Exercises 3 and 4, use the table at the right. It shows the number of species at 37 major U.S. public zoological parks.

Number of Species				
200	700	290	600	681
300	643	350	794	400
360	600	134	200	800
305	384	500	330	250
530	715	303	200	475
465	340	347	300	708
184	800	375	350	450
337	221			

3. Make a histogram of the data. Use intervals that begin at 100, 200, 300, 400, 500, 600, 700, and 800.

Animal Species in Zoos

4. Which interval has the largest frequency?

HEALTH For Exercises 5 and 6, use the graph at the right.

5. What does each bar represent?

6. Determine whether the graph is a bar graph or a histogram. Explain how you know.

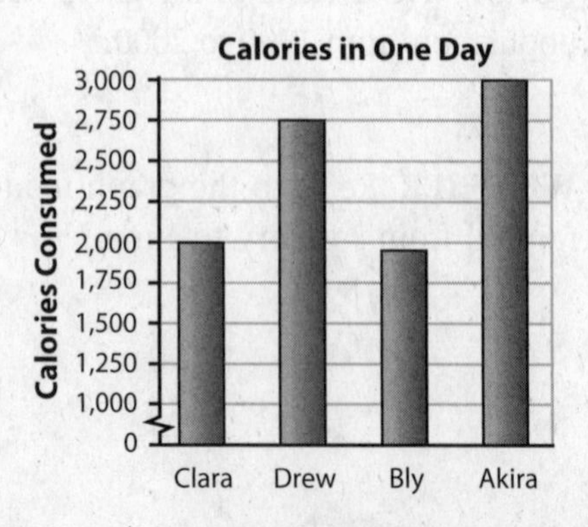

Name______________________________ Date_______ Period_______

Math Skills Study Guide

Stem-and-Leaf Plots

In a **stem-and-leaf plot,** the data are organized from least to greatest. The digits of the least place value usually form the **leaves,** and the next place value digits form the **stems.**

EXAMPLE 1 **Make a stem-and-leaf plot of the data below. Then find the range, median, and mode of the data.**
42, 45, 37, 46, 35, 49, 47, 35, 45, 63, 45

Order the data from least to greatest.
35, 35, 37, 42, 45, 45, 45, 46, 47, 49, 63

The least value is 35, and the greatest value is 63. So, the tens digits form the stems, and the ones digits form the leaves.

Stem	Leaf
3	5 5 7
4	2 5 5 5 6 7 9
5	
6	3

6|3 = 63

range: greatest value − least value = 63 − 35 or 28
median: middle value, or 45
mode: most frequent value, or 45

EXERCISES

Make a stem-and-leaf plot for each set of data. Then find the range, median, and mode of the data.

1. 15, 25, 16, 28, 1, 27, 16, 19, 28

2. 1, 2, 3, 2, 3, 1, 4, 2, 5, 7, 12, 11, 11, 3, 10

3. 3, 5, 1, 17, 11, 45, 17

4. 4, 7, 10, 5, 8, 12, 7, 6

Name ________________________ Date ______ Period ______

Math Skills Study Guide

Stem-and-Leaf Plots

Make a stem-and-leaf plot for each set of data.

1. 23, 36, 25, 13, 24, 25, 32, 33, 17, 26, 24

2. 3, 4, 6, 17, 12, 5, 17, 4, 26, 17, 18, 21, 16, 15, 20

3. 26, 27, 23, 23, 24, 26, 31, 45, 33, 32, 41 40, 21, 20

4. 347, 334, 346, 330, 348, 347, 359, 344, 357

HOT DOGS For Exercises 5–7, use the stem-and-leaf plot at the right that shows the number of hot dogs eaten during a contest.

Stem	Leaf
0	8 8 9
1	1 2 2 4 7 7 7
2	1 1 2

2 | 1 = 21

5. How many contestants are represented on the stem-and-leaf plot?

6. How many hot dogs did they eat all together?

7. What is the range of the number of hot dogs eaten?

8. Find the median and mode of the data.

Determine the mean, median, and mode of the data shown in each stem-and-leaf plot.

9.

Stem	Leaf
0	1 2 2 3
1	3 4 5 5
2	0 0 0 1 3

2 | 0 = 20

10.

Stem	Leaf
2	0 0 0 2 3 5 7
3	1 2
4	0

4 | 0 = 40

11.

Stem	Leaf
22	1 1 2 7
23	3 3 9
24	0 6 8

24 | 0 = 240

Name _______________________________ Date _______ Period _______

Spiral Review

Unit 1, Lesson 1

1. In planning a post-prom party, the senior class officers at Kennedy High School get a price quotation from a local athletic club. There would be a basic charge of $450 for the facility plus $10 per student for food and drinks.

   a. Display the (*number of students, cost*) values in the following table.

Number of Students	0	50	100	150	200	250	300
Cost							

   b. The class officers decide to charge each student $15 to attend the party. What income will they get if 250 students buy tickets?

   c. Display (*number of tickets sold, income*) values in a table and in a graph.

Number of Tickets	0	50	100	150	200	250	300
Income							

   d. The senior class president proposed the rule $I = 5n - 450$ for the relationship between number of tickets sold, n, and the income, I. Does this rule represent the pattern in the table and graph? Explain your reasoning.

   e. Write and solve an equation to find the number of tickets sold that will allow the senior class to break even on the event. Explain how this solution is shown on the graph.

Name ____________________ Date ________ Period ________

Spiral Review

Unit 1, Lesson 2

1. Imagine folding a square piece of paper in half, then in half again, and then in half again, and so on. The fold marks at each stage divide the original square into a number of smaller regions.

   a. Complete the table showing the number of regions for n folds.

n folds	1	2	3	4	5
R regions					

   b. Write a NOW-NEXT equation showing how to calculate the number of smaller regions at any stage of the folding process.

   c. Make a graph showing the relationship between the number of folds and the number of regions. Describe the pattern in the values.

   d. Find the number of regions for 8 folds.

2. According to the U.S. Census Bureau, the population of the United States in July of 2006 was approximately 299 million and growing at a rate of about 1.05% per year.

   a. Estimate the U.S. population in July of 2007, 2008, and 2009.

   b. Write a NOW-NEXT equation that can be used to project the U.S. population any number of years into the future.

   c. Estimate the year when the U.S. population will first exceed 350 million.

Name ______________________ Date ________ Period ________

Spiral Review

Unit 1, Lesson 3

1. Suppose a provider of local telephone service offers to charge only $15 per month plus $0.15 per call.

   a. What will the monthly bill be if 45 calls are made?

   b. What equation shows how to calculate the monthly bill, y, as a function of the number of calls made, x?

   c. How many calls must have been made if one monthly bill was $30?

2. The vendors at a baseball stadium are paid $20 per game and 10% of the value of the food and drinks they sell.

   a. Explain how the rule $P = 20 + 0.1d$ shows how the vendor pay depends on the dollar value of food sold by that vendor.

   b. Determine the game pay for a vendor who sells $350 worth of food and drinks.

   c. How much food and drink must a vendor sell to earn a game pay of $75?

3. Without use of your graphing calculator or computer software, match the following four rules to the graph sketches below. Explain your reasoning in each case. Scales on the axes are 1.

   a. $y = 2x^2 - 4$

   b. $y = 2^x$

   c. $y = 2x - 4$

   d. $y = \frac{2}{x}$

I

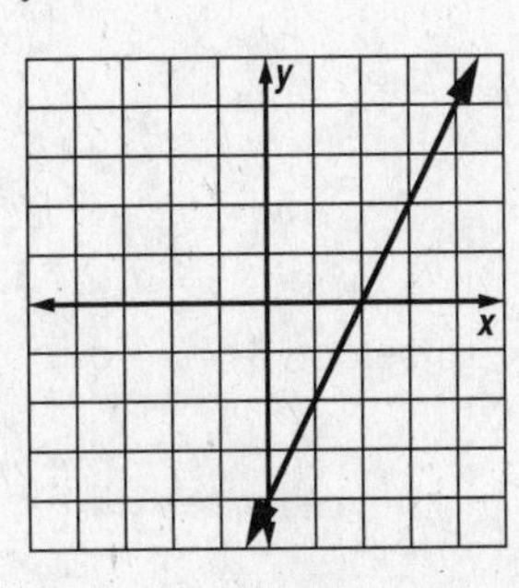

II

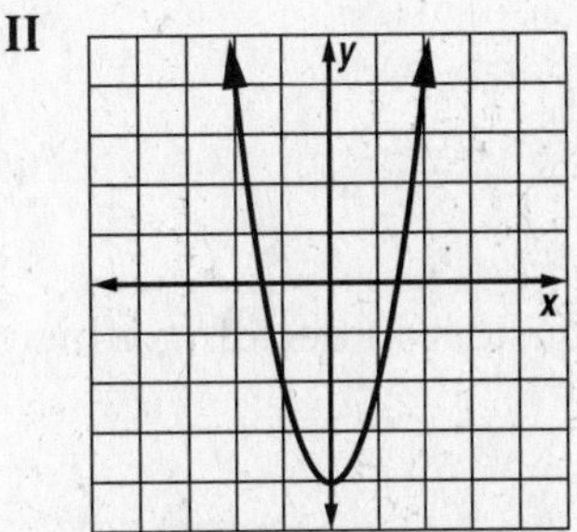

III

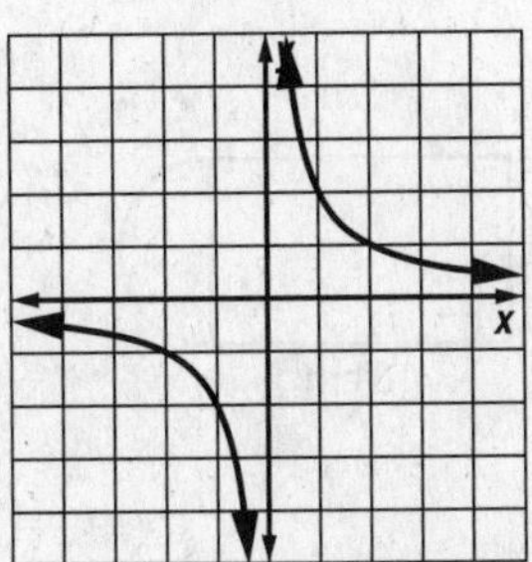

IV

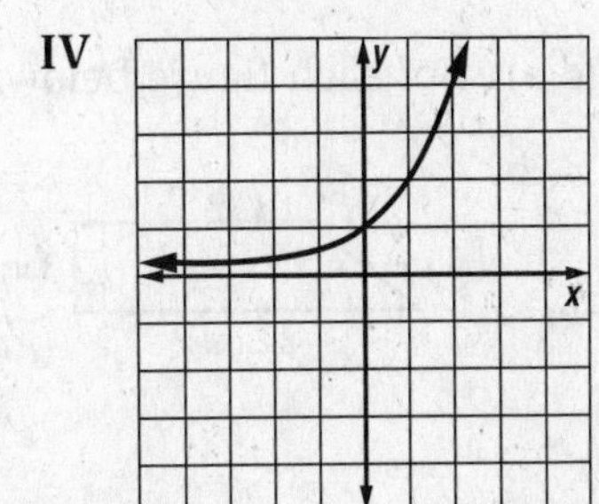

Name ______________________________ Date ________ Period ________

Spiral Review

Unit 1, Lessons 1, 2, and 3

1 The Rodriguez family is taking a trip to visit an amusement park that is 200 miles away. They want to arrive at the amusement park as soon as it opens at 9:00 AM.

a. If the average speed during the trip is 60 miles per hour, what time will they need to leave home in order to be at the park when it opens? When would they need to leave if the average speed is 40 miles per hour?

b. Make a table showing how the *drive time* changes as the *average speed* changes from 35 miles per hour to 70 miles per hour, in increments of 5 miles per hour.

c. Describe the pattern of change shown in your table and graph.

2 Neil has accepted a new job that will pay an annual salary of $28,000 for the first year. He has the option of a 5% annual raise or a fixed annual raise of $1,500.

a. For each option, write a NOW-NEXT equation that shows how Neil's salary will increase from year to year.

b. For each option, make a table showing how Neil's salary increases over a period of 10 years.

c. What should Neil consider in deciding which alternative to select?

3 Find the perimeter and area of each figure below.

a.

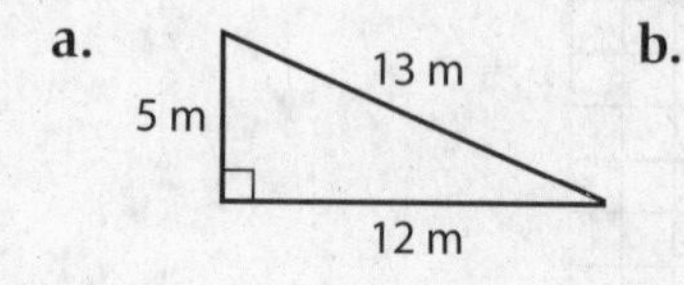

b.

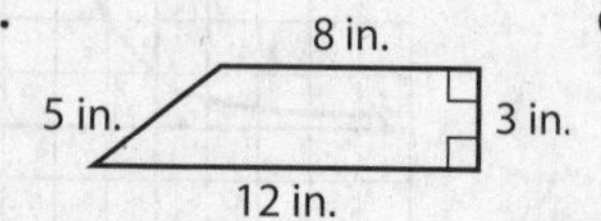

c.

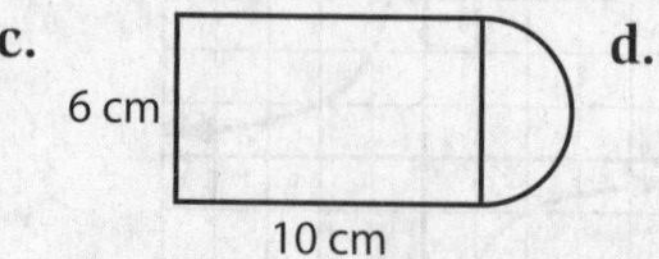

d.

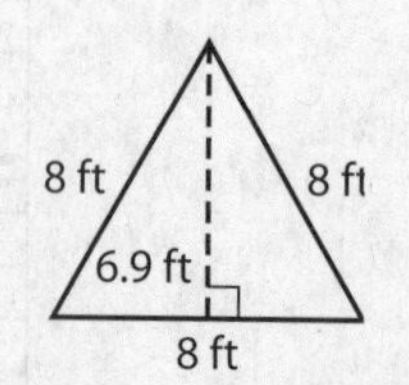

Name ______________________ Date ________ Period ________

Spiral Review

Unit 2, Lesson 1

1 On the first three tests in a marking period Paula has scores of 85, 90, and 75.

a. What is the lowest possible score Paula can get on the fourth and final unit test in order to have a mean of at least 85 for the marking period?

b. What will her median score be for the marking period if she gets the lowest possible score in Part a?

c. What will her range of scores be for the marking period of she gets the lowest possible score in Part a?

2 Find the range, mean, and median of the following set of numbers:

6.72, 5.803, 3.5, 7, 8.07

3 Consider the following list of high temperatures (°F) for the first two weeks of October in a city in Maryland.

65, 72, 60, 64, 75, 59, 71, 63, 60, 67, 72, 84, 86, 63

a. Make a histogram of the data.

b. Describe the shape of the distribution.

c. Calculate the mean high temperature for this time period.

d. Calculate the median high temperature for this time period.

Name ________________________________ Date ________ Period ________

Spiral Review

Unit 2, Lesson 1

1. On the first five quizzes of a marking period, a student has a mean of 6 on a scale of 0 to 10. Find the student's new mean quiz score if, on the next 10-point quiz, the student earned

   a. A score of 10 **b.** A score of 6 **c.** A score of 4

2. International Falls, Minnesota is often the coldest spot in the lower 48 states during the winter season. During one week in January, the low temperatures in Fahrenheit were: $-5°$, $-15°$, $-20°$, $-35°$, $-4°$, $6°$ and $10°$.

   a. What was the mean and median low temperature for that week?

   b. What was the range of low temperatures for that week?

3. To compare gasoline prices in two neighboring states, students in a *Core-Plus Mathematics* class collected data from 10 service stations in each state. The costs per gallon (in dollars) for regular unleaded gasoline were:

   State 1: 2.879, 2.839, 2.889, 2.919, 2.849, 2.739, 2.829, 2.869, 2.799, 2.859

   State 2: 2.789, 2.769, 2.799, 2.929, 2.859, 2.829, 2.819, 2.789, 3.089, 2.999

   a. What are the mean, median, and range of prices in the sample of stations from each state?

   b. Which state seems to have the lower gasoline prices, and what data summaries best support your conclusion?

Review (Unit 1, Lesson 3)

4. Louis weighs 280 pounds and has just joined a weight loss program that guarantees he will lose 10 pounds per month until he reaches his target weight. Write an expression for Louis' weight after m months.

Name ______________________________ Date ________ Period ________

Spiral Review

Unit 2, Lesson 2

1. Here are the ratings given by two different judges in a gymnastics contest. Possible scores range from 0 to 6.0.

Judge 1: 3.6, 3.6, 4.4, 4.5, 4.5, 4.6, 4.6, 4.8, 4.9, 5.2, 5.3, 5.3, 5.4, 5.5, 5.7, 5.8, 5.9, 5.9, 6.0

Judge 2: 3.4, 3.7, 3.8, 3.9, 4.2, 4.3, 4.7, 4.7, 4.8, 4.8, 5.3, 5.4, 5.4, 5.5, 5.6, 5.7, 5.9, 6.0, 6.0

a. Calculate the summary statistics needed to construct box plots of the data and draw those plots.

b. Calculate the mean and the standard deviation for each judge's scores.

c. Based on the various summary statistics you've calculated and the box plots you've drawn, what conclusion would you reach on the question of whether the two judges give similar or different ratings overall to a group of skaters?

2. The following data show average class sizes in a sample of 10 states.

State	AL	AZ	DE	IN	ME	MI	NM	SC	WA	WI
Average Class Size	16.6	19.7	16.6	17.3	13.7	19.1	16.7	15.7	20.2	16.1

Source: *The World Almanac and Book of Facts, 1999.* Copyright 1998 World Almanac Education Group. All rights reserved.

a. Calculate the mean and median class size of the sample.

b. Calculate the standard deviation of class size in the sample and explain what the statistic tells about the average class size that the measures of center don't reveal.

3. Suppose the class mean for 20 students on a 10-point test is 7.5, the median is 8, and the range is 3 points. What will happen to these summary statistics if the teacher:

a. Adds 1 point to every student's score?

b. Multiplies each score by 10?

c. Includes one new score of 6 for a student who took a makeup test?

Name ______________________________ Date ________ Period ________

Spiral Review

Unit 2, Lessons 1 & 2

1. The following numbers are averaged freshman high school graduation rates (% of freshman class graduating in 4 years) for 48 of the United States, plus the District of Columbia, in 2003–2004. (Statistics for New York and Wisconsin were not available.)

57.4, 60.6, 61.2, 62.7, 65.0, 66.1, 66.4, 66.8, 67.0, 67.2, 68.2, 69.4, 71.4, 72.5, 72.6, 72.9, 73.0, 73.5, 73.9, 74.2, 74.6, 75.9, 76.0, 76.7, 76.8, 76.9, 77.0, 77.6, 77.9, 78.7, 78.7, 79.3, 79.3, 79.5, 80.3, 80.4, 80.4, 80.7, 81.3, 81.5, 82.2, 83.0, 83.7, 84.7, 85.4, 85.8, 86.1, 86.3, 87.6

Source: *National Center for Education Statistics, "The Averaged Freshman Graduation Rate for Public High Schools form the Common Core of Data: School Years 2002-03 and 2003-04" http://nces.ed.gov/pubs2006/2006606.pdf*

a. Display the data in a histogram, grouping the data in intervals of 5 percentage points.

b. Describe the shape of the distribution. Based on this shape, how do you expect the mean and median rates to compare?

c. Calculate the mean and median graduation rates.

d. Calculate the five-number summary for the data. Explain what each number tells you about the graduation rates in the United States for the school year 2003–2004.

e. Calculate the standard deviation for the data. Explain what this number tells you about the graduation rates in the United States for the school year 2003–04.

f. Diploma counts were missing for New York and Wisconsin. Using estimated counts from these two states, the mean graduation rate for the United States is 74.3%. Give two possible graduation rates for these states that would produce a mean of 74.3%.

Name ______________________ Date ________ Period ________

Spiral Review

Unit 3, Lesson 1

1 Determine the slope and y-intercept of each linear function.

a. $y = -\frac{7}{2} + \frac{5}{2}x$

b. The line with the graph below.

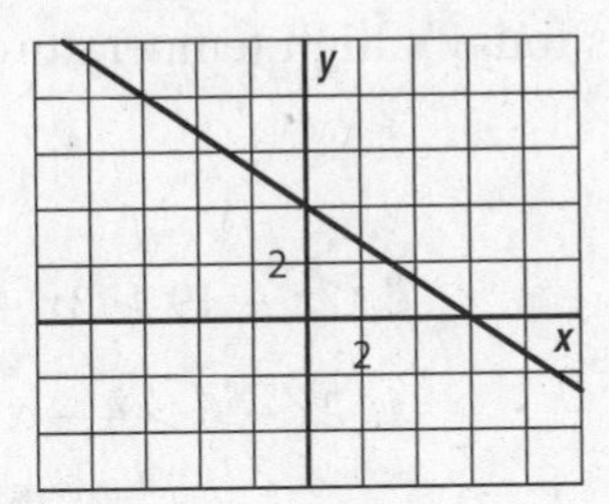

c. The line with the table below.

x	y
−4	−21
−2	−16
0	−16
2	−6

d. The line given by NEXT = NOW − 2.1 (start at 4)

2 Sketch the graphs of the following equations. Use a separate piece of graph paper.

a. $y = 2x - 3$

b. $y = -\frac{3}{5}x + 1$

c. $y = -5$

d. $y = 4 + x$

e. $x = 2$

f. $y = 3 - \frac{3}{4}x$

3 Write equations for the lines satisfying these conditions.

a. Passes through the points (2, 5) and (4, −3)

b. Is parallel to the line with equation $y = 10 + 4x$ and contains the point (5, −8)

4 The students in a high school class who baby-sat over spring break collected data on the number of hours they baby-sat and the amount they charged per hour. Their data appear in the following table.

Charge Per Hour ($)	7	2	0	2	6	3
Number of Hours Worked	3	8	15	10	5	9

Find the equation of the regression line modeling the relationship between charge and number of hours worked. Graph that line on a plot of the data.

Name ______________________________ Date ________ Period ________

Spiral Review

Unit 3, Lesson 2

1. The number of visitors V to a swimming pool varies with the day's high temperature in degrees Fahrenheit, T, according to the following equation:

$$V = 150 + 25(T - 80)$$

If 500 people are predicted to visit the pool today, what is today's high temperature?

2. Solve each of the following inequalities.

a. $3x + 12 < 54$ **b.** $7x - 19 > 30$ **c.** $7x - 19 \geq 3x + 13$

d. $13 - 4x < 25$ **e.** $\frac{1}{5}x + 3 \leq 9$ **f.** $2x - 6 > 4 - x - 3$

3. Solve the following systems of equations. Show your work.

a. $3x + 2y = 5$, $-2x + y = -1$

b. $y = 4 - 7x$, $y = 13 + 5x$

c. $3x + y = 3$, $x - 5y = 9$

4. A high school choir is selling kits for plain cheese pizzas and pepperoni pizzas to raise money for their spring trip. One student submitted an order for 23 pizza kits with no indication of how many orders were for cheese and how many were for pepperoni. The student also submitted a check for $187. If the cheese pizza kits sell for $7 and the pepperoni pizza kits sell for $9, how many of each type of kit were ordered?

Review (Unit 2, Lesson 2)

5. The following data show sales by two concession stands at a baseball stadium during the first 11 days of the season.

Stand 1 Sales ($): 250 190 200 185 210 120 175 140 125 180 110

Stand 2 Sales ($): 225 160 180 200 240 110 150 180 110 140 90

Calculate the summary statistics needed and draw box plots comparing the sales data from the two stands.

Name ______________________ Date ________ Period ________

Spiral Review

Unit 3, Lesson 3

1. If $y = 4(-3x + 5) - 9$, find y if:

   a. $x = 10$ **b.** $x = -10$ **c.** $x = 0.5$

   d. $x = 0$ **e.** $x = -1$ **f.** $x = \frac{2}{3}$

2. Solve each linear equation.

   a. $3x - 12 = 24$ **b.** $6x - 17 = 20 - 9x$ **c.** $10 - (5 - 2x) = 7$

   d. $-4(2x + 8) = 3(4 - x)$ **e.** $x = 2x - 9$ **f.** $\frac{1}{2}x + 2 = \frac{5}{2}x - 10$

3. The school Booster Club is planning to sell state championship T-shirts. They expect the following expenses and income.

   Expenses: \$50 art-screen fee, \$5.75 per shirt

   Income: \$10 per shirt

   a. Write an expression for the cost of n shirts.

   b. Write an expression for the income earned from the sale of n shirts.

   c. Write two equivalent expressions for the profit earned from the sale of n shirts.

   d. What is the minimum number of shirts that must be sold in order not to lose money?

4. In each case below, a student has made an error in attempting to write an equivalent expression. Spot the reasoning error and write an explanation to help clear up the problem for the student who made the error.

   a. $5(2x + 3) = 10x + 3$ **b.** $3x - 7x = 4x$

   c. $8 -2(3x - 4) = -6x$ **d.** $2 + 3(x + 1) = 5x + 5$

5. Write the function rule $y = 3(x + 2) - 5$ in three equivalent forms.

Name ______________________________ Date ________ Period ________

Spiral Review

Unit 3, Lessons 1, 2, & 3

1 Two lines are graphed on the coordinate axes to the right. The scale on each axis is 1.

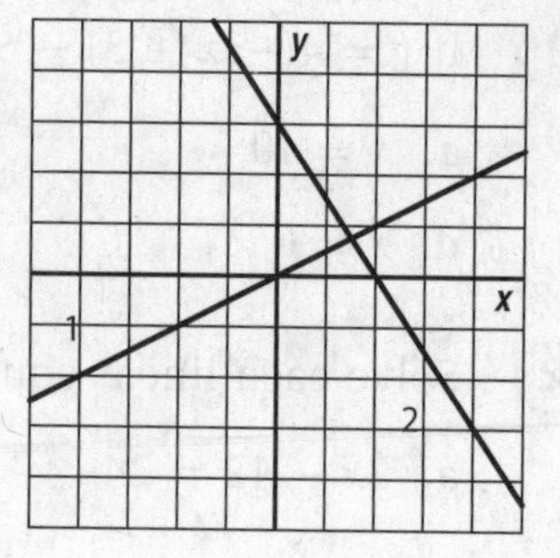

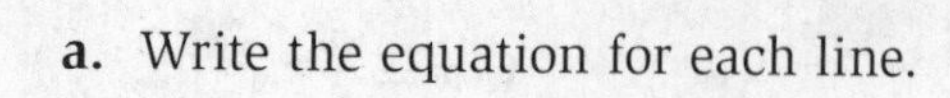

a. Write the equation for each line.

b. Determine the exact coordinates of the point of intersection of the two lines.

2 The first 4 stages of a pattern are shown below.

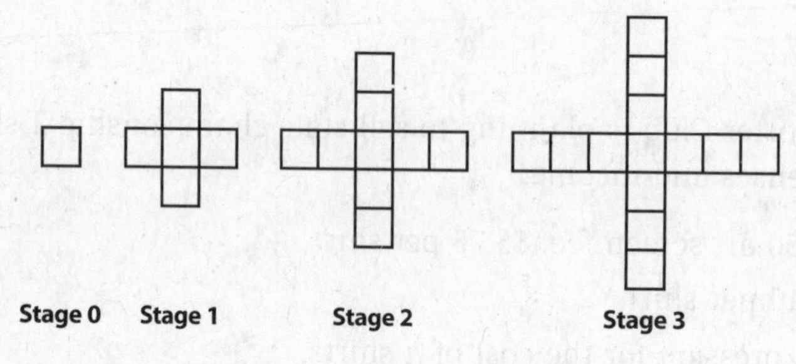

a. Make a table showing the number of square tiles used for each of the first eight stages of the pattern. Make a graph of the data.

b. Write an equation using NOW and NEXT that describes the number of tiles in one stage given the number of tiles in the previous stage.

c. Write an equation to find the number of square tiles, T, used at the nth stage of the pattern. Draw this line on your graph.

3 Solve each of the following for n.

a. $3n + 14 = 2 - n$

b. $2(5n - 1) = 7(n + 1)$

c. $3n + 4 < 1$

d. $\frac{(4 - n)}{2 + 5} \geq 9$

4 Solve the following system of equations in three different ways.

$y = -\frac{2}{3}x + 4$ and $y = \frac{1}{2}x + 2$

Name ______________________________ Date ________ Period ________

Spiral Review

Unit 3, Lessons 1, 2, & 3

1. Andrea's parents have rented a suite at a Phoenix Mercury basketball game for her birthday party. The suite costs $180 to rent. In addition, they must buy $12 tickets to the game for each person. Write and solve equations or inequalities to answer the following questions.

   a. How much will it cost for 8 people to be at the party?

   b. If Andrea's parents paid $240, how many people were at the party?

   c. How many people can be at the party if the cost must be less than $320?

2. The graph of a linear function contains the points (3, –2) and (–6, 1). Write a linear function rule for the line.

3. Write the formula for perimeter P of a rectangle with length L and width W in two equivalent forms and explain how you know the forms are equivalent.

4. Solve each of the following equations by reasoning without the use of tables or graphs. Then check your work using a calculator-based solution strategy.

   a. $5.5x + 23 = -15$

   b. $9 - 4x = 17$

   c. $5 + 3x = 27 - 5x$

   d. $4(3x + 7) = 75$

   e. $3(4x - 7) = 8x + 14$

   f. $-4(3 - x) = 11(x + 2)$

Review (Unit 2, Lesson 2)

5. Suppose the scores on a 100-point test for a class of 20 students have mean 75%, median 80%, and range 40 (from 55% to 95%). How will the mean, median, and range change (if at all) if the teacher:

   a. increases each student score by 5 points?

   b. Divides each score by 10?

Name ______________________________ Date ________ Period ________

Spiral Review

Unit 3, Lessons 1, 2, & 3

1 The following diagram shows how a rigid roof brace can be constructed by connecting short bars in a triangulated pattern.

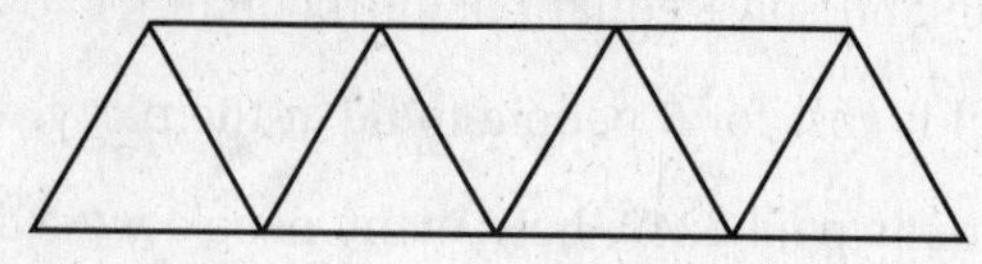

a. Make a table showing the number of bars needed to make such a brace so that the bottom side has a length of 1, 2, 3, 4, or 5 bars.

b. Write a rule showing how the number of bars B required for a brace depends on the length of the bottom side n.

2 Determine the slope and y–intercept of the graph of each linear function below.

a. The function with the graph shown below.

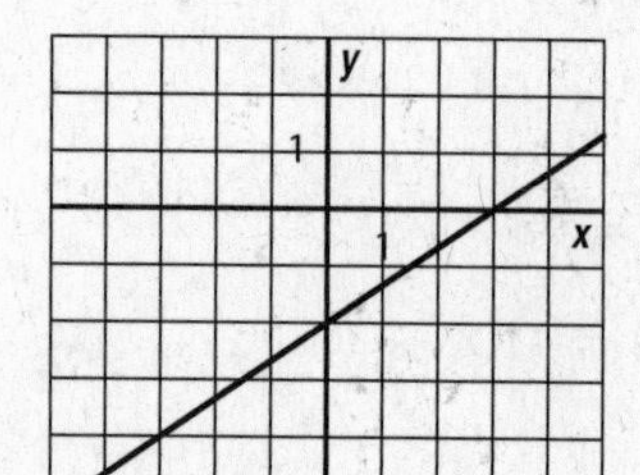

b. The function with the table below.

x	y
−3	3
−1	2.3333
1	1.6667
3	1

3 Solve each inequality for the indicated variable.

a. $-12 + 11m \leq 54$

b. $7(r + 8) < 3(r + 12)$

4 Without using a graphing calculator, graph each of the following rules. Use a separate piece of graph paper.

a. $y = x$

b. $y = 5$

c. $y = 10 - 1.5x$

Name ______________________________ Date ______ Period ______

Spiral Review

Unit 4, Lesson 1

1. Determine whether each graph has an Euler circuit.

a.

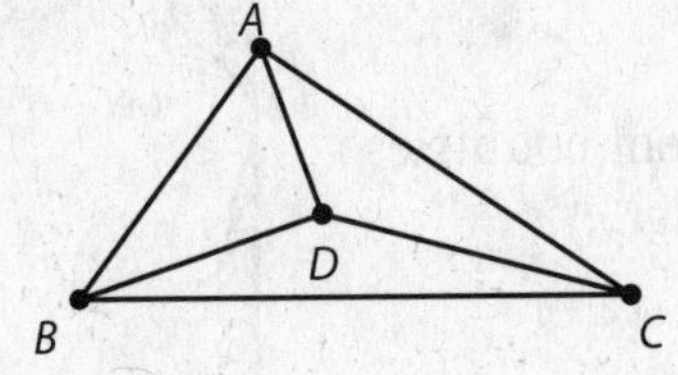

b.

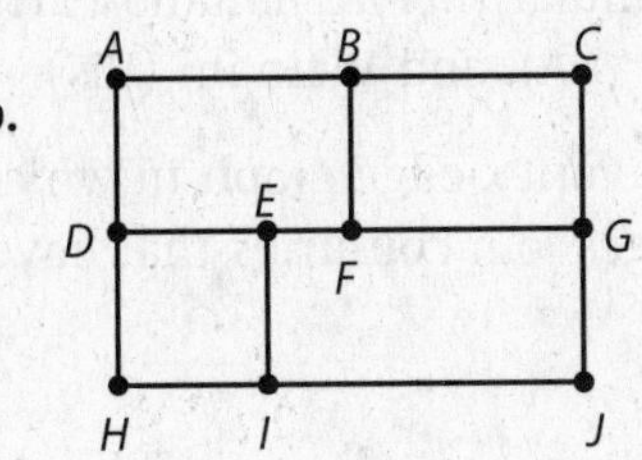

2. Which of the following diagrams is an Euler circuit? Explain your reasoning.

a.

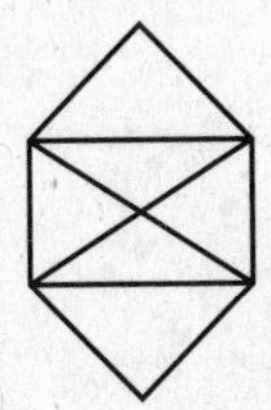

b.

3. Consider the following vertex-edge graphs.

I

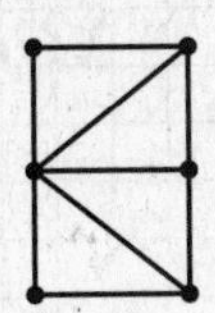

II

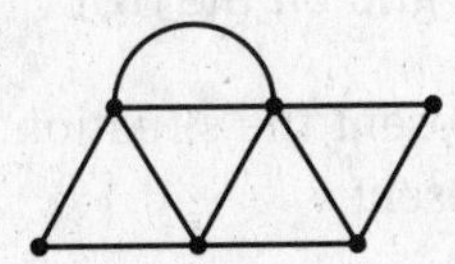

III

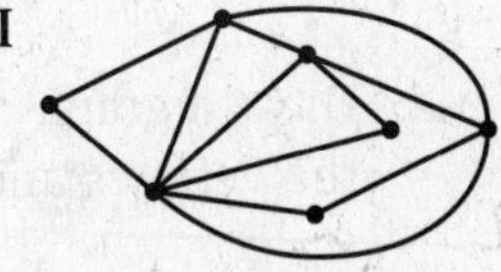

Which of the graphs contain an Euler circuit? Explain your answer.

Name ______________________ Date ________ Period ________

Spiral Review

Unit 4, Lessons 1 & 2

1 The sketch at the right shows a map of six countries in Southeast Asia—Myanmar (MY), Thailand (TH), Malaysia (MA), Cambodia (CA), Laos (LA), and Vietnam (VN).

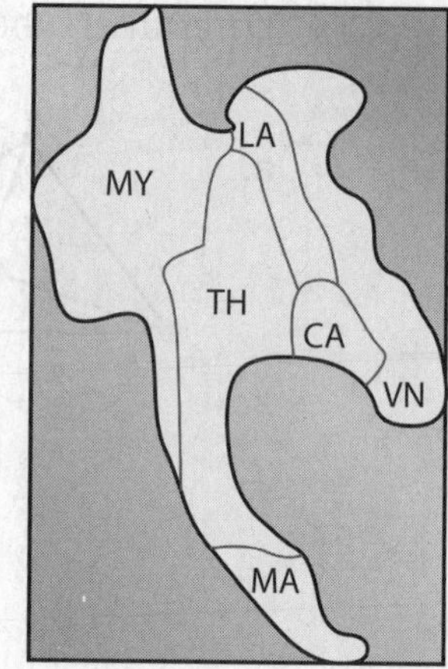

a. Draw a vertex-edge graph in which vertices represent countries and edges join countries that have a common border.

b. Determine whether there is either an Euler circuit or an Euler path for the graph in part a. Explain what such a circuit or path would mean to someone traveling in the six countries.

c. Determine the minimum number of colors needed to color the map so that no countries with a common border have the same color.

2 Seven radio stations are planning to start broadcasting in the same region of the country. Stations within 500 miles of each other on the same frequency will interfere with one another. The locations of the seven stations are shown on the grid on the right.

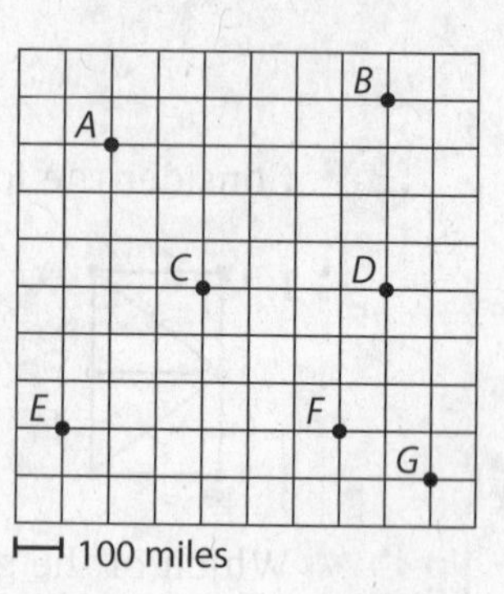

a. Draw a graph model to represent the situation. Indicate what the vertices and edges represent.

b. Use graph coloring to assign as few frequencies as possible to the seven radio stations.

Name ______________________________ Date ________ Period ________

Spiral Review

Unit 5, Lesson 1

1 Suppose a radio station has about 5,000 listeners during the morning rush hour period. A new station manager sets a goal of increasing that number by 10% every month.

a. Make a table giving the necessary number of listeners to meet the goal for each of the next five months.

b. Write a NOW-NEXT rule to calculate the necessary number of listeners to meet the goal in *any* future month. Then write a rule relating the number of listeners, L, after any number of months, m.

2 Without using a graphing calculator, sketch graphs of these "$y = \ldots$" rules. Use a separate piece of graph paper.

a. $y = 4(1.5^x)$ **b.** $y = 3(2.5^x)$

Review (Unit 3, Lesson 2)

3 Cindy is investigating how fast a particular bee population will grow under controlled conditions. She began her experiment with 2 bees. The next month she counted 10 bees.

a. Write a rule for the number of bees B after m months if the pattern of growth is linear.

b. Write a rule for the number of bees B after m months if the pattern of growth is exponential.

c. How many months will it take for the number of bees to reach 200 assuming linear growth? Assuming exponential growth?

Review (Unit 2, Lesson 2)

4 The box plot at the right shows the price in dollars of 20 models of cordless phones. Give the five-number summary of the data. Then explain what the box-plot tells you about the prices of cordless phones.

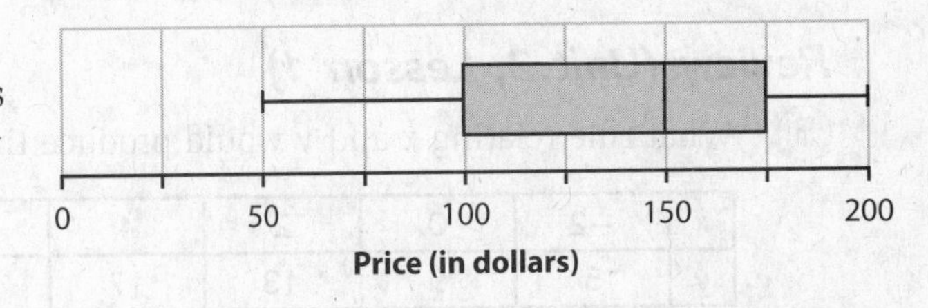

Name ______________________________ Date ________ Period ________

Spiral Review

Unit 5, Lesson 1

1 Write the NOW-NEXT and $y = \ldots$ rules that match each of the tables below.

a.

x	0	1	2	3	4	5
y	1.5	3	6	12	24	48

b.

x	0	1	2	3	4	5
y	2	6	18	54	162	486

c.

x	0	1	2	3	4	5
y	4	5	6.25	7.8125	9.765625	12.20703125

2 Write each expression in a shorter form using exponents.

a. $x^3x^4yy^5$ **b.** $x^7y^3(xy)^5$ **c.** $(3x^4)^2$

3 Aaron invests $500 in a certain investment portfolio, which is expected to have a growth rate of 8% per year.

a. Make a table showing how much Aaron will have each year if he invests for 5 years.

b. Write a $y = \ldots$ rule showing the relationship between years, n, and dollars, D.

c. If Aaron does not invest any more money, how long will it take before he can expect to have at least $1,000 in the account?

4 Find values for x and y that will make these equations true statements.

a. $(5^3)^4 = 5^x$ **b.** $3^54^2 \times 3^34 = 3^x4^y$ **c.** $(2n^3)^4 = 2^xn^y$ **d.** $(5^2)^x = 5^{14}$

e. $(3.8^2)^3 = x^6$ **f.** $(42)^4 = 6^x \cdot 7^y$ **g.** $(r^3s^4)(r^2s) = r^xs^y$ **h.** $(h^3j^2)^x \cdot (h^2j^3) = h^8j^7$

Review (Unit 3, Lesson 1)

5 What rule relating x and y would produce the (x, y) pairs in the table below?

x	−2	0	2	4	6
y	5	9	13	17	21

Name ______________________ Date ________ Period ________

Spiral Review

Unit 5, Lesson 2

1 Write a NOW-NEXT rule and a $y = \ldots$ rule to match each graph below.

a.

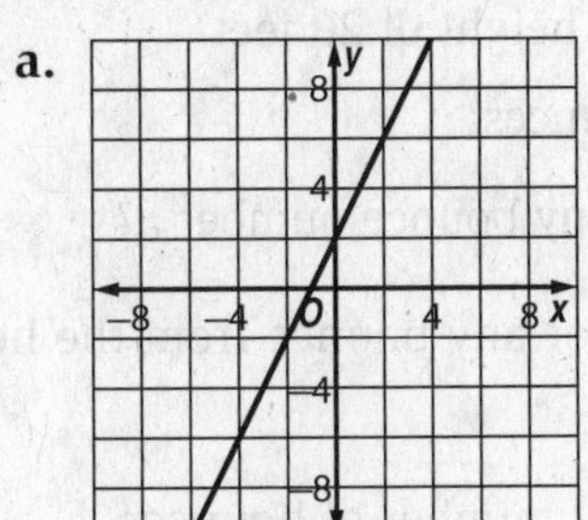

b.

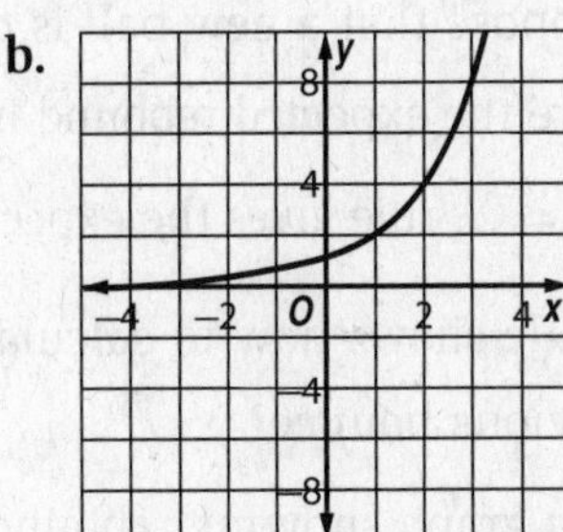

2 Suppose a hospital patient receives an injection of medication that metabolizes in the blood according to the rule $M = 200(0.8)^t$ (with M in milligrams and t in hours).

a. What do the values 200 and 0.8 tell about the action of the medicine in the bloodstream?

b. What is the value of M when $t = 3.5$, and what does it tell about medicine action?

c. Sketch a graph of the (*time, medication*) relation.

3 Write each of the following expressions in shorter equivalent form using exponents.

a. $a \cdot a \cdot b \cdot b \cdot b \cdot (2.3) \cdot (2.3)$ **b.** $2 \cdot 2^3 \cdot a^2 \cdot a^4$ **c.** $\dfrac{(a^3b^4c)}{(ab^3c^5)}$

4 If a new truck costs \$25,000, its trade-in value will decrease by about 15% each year after purchase.

Write rules in NOW-NEXT and $y = \ldots$ form showing how to calculate the truck's trade-in value for any number of years after purchase.

Name ______________________________ Date ______ Period ________

Spiral Review

Unit 5, Lesson 2

1 If a tennis ball is dropped onto a hard surface, it should rebound to about 50% of its drop height. Suppose that a new ball is dropped from an initial height of 20 feet.

a. What are the expected rebound heights for the first 3 bounces?

b. What $y = \ldots$ rule gives the expected rebound height for any bounce number n?

c. What rule shows how to calculate the rebound height for any bounce from the height of the previous bounce?

d. Sketch a graph showing rebound height as a function of number of bounces.

2 Evaluate each expression.

a. $\left(\frac{2}{3}\right)^{-1}$ **b.** $8^{\frac{1}{3}}$ **c.** 3^{-2}

d. 25^0 **e.** $\left(\frac{1}{4}\right)^{-2}$ **f.** $\sqrt{\frac{36}{25}}$

3 Rewrite each of the following expressions in a simpler form.

a. 3^0 **b.** $49^{\frac{1}{2}}$ **c.** n^3n^5

d. $(2a^3b)^4$ **e.** $\frac{x^8y^3}{x^2y}$ **f.** $\frac{15x^6y^7}{5x^2y^4}$

Review (Unit 3, Lesson 1)

4 Without use of a graphing calculator, sketch graphs of these rules. Use a separate sheet of graph paper.

a. $y = 0.5x$ **b.** $y = 3 + 0.5x$

c. $y = 3 - 0.5x$ **d.** $y = 0.5 - 3x$

e. $y = 3(0.5)^x$ **f.** $y = 0.5(3)^x$

Review (Unit 2, Lesson 1)

5 Angela did yard work for her neighbors over the summer. Her weekly earnings were $40, $20, $23, $38, $13, $34, $100, $25, $10, $5, and $31.

Calculate and compare the mean and the median weekly earnings for Angela. Why does it make sense that the mean is greater than the median?

Name ______________________________ Date ________ Period ________

Spiral Review

Unit 6, Lesson 1

1. Find the missing dimensions of the television screens pictured below.

a.

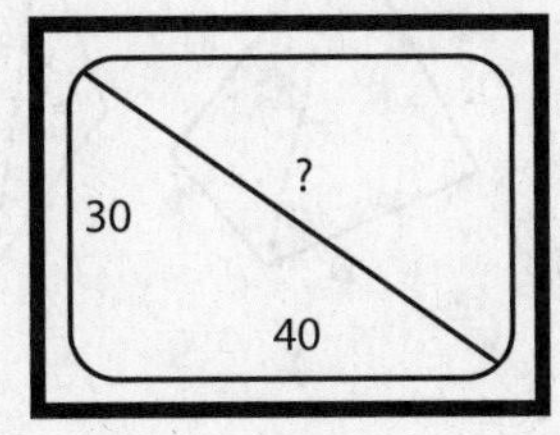

b.

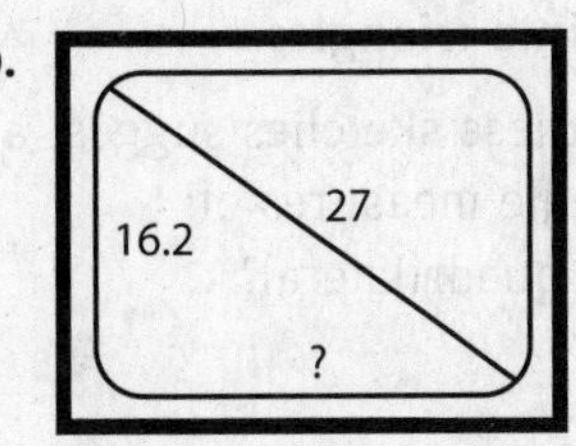

2. Are the two triangles shown below congruent? Explain how you know.

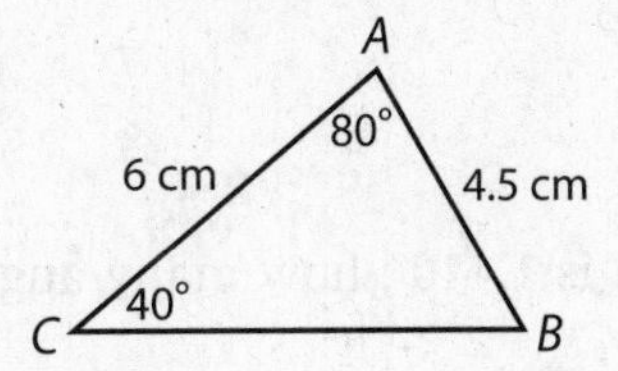

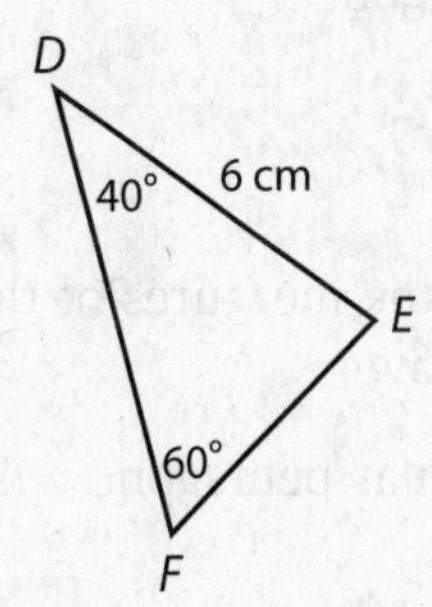

3. The figure at the right is a parallelogram. What conclusions can you draw about the measurements in degrees of the labeled angles?

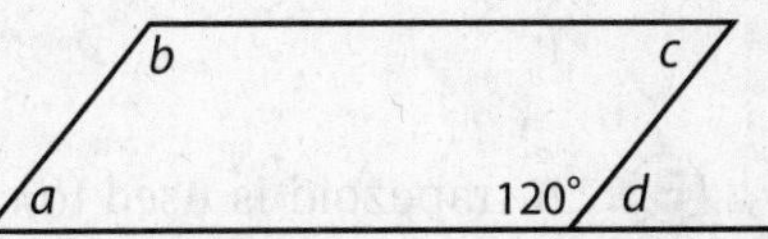

4. If the angles of a triangle are related as indicated by the following sketch, what are the measures in degrees of those angles?

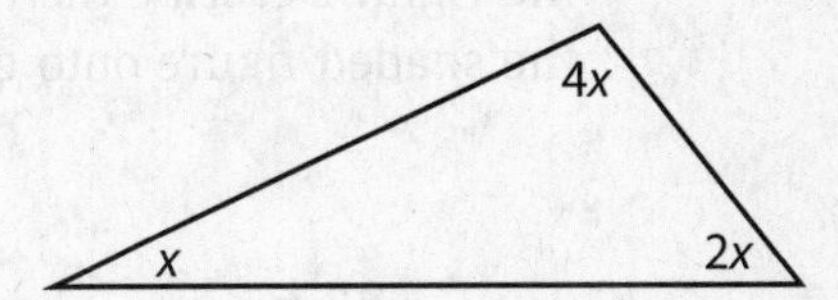

5. Sketch the pair of quadrilaterals below. Describe their similarities and their differences.

Parallelogram and rhombus

Name ______________________________ Date ________ Period ________

Spiral Review

Unit 6, Lesson 2

1 The sketches below show how convex polygons can be subdivided into triangles.

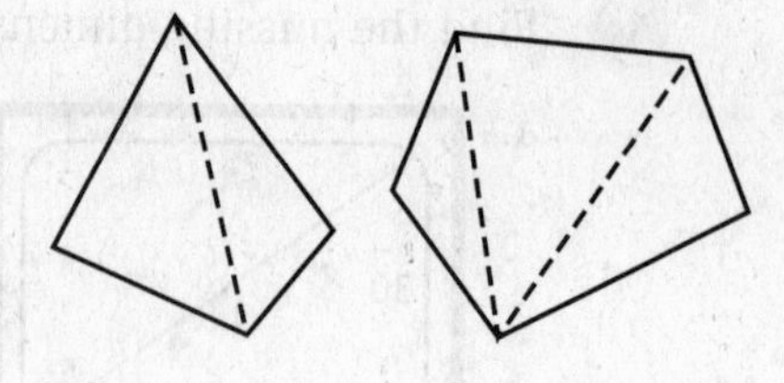

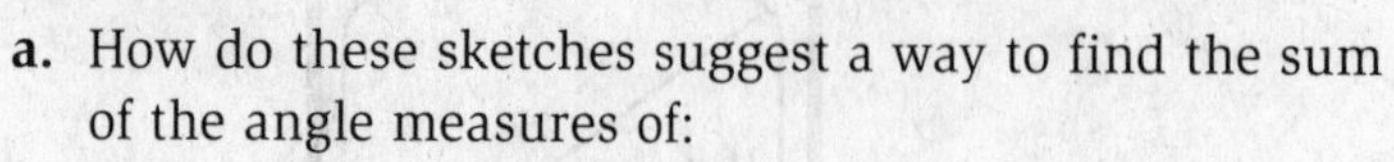

a. How do these sketches suggest a way to find the sum of the angle measures of:

(i) any quadrilateral?

(ii) any pentagon?

(iii) any hexagon?

(iv) any n-gon?

b. If the sum of the measures of the angles of a polygon is 1,440°, how many angles does the polygon have?

2 Explain why regular pentagons will not tessellate.

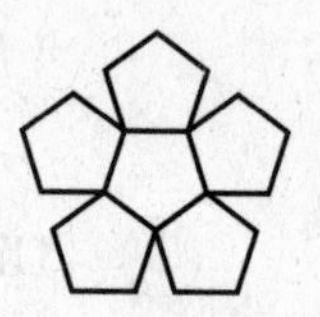

3 A trapezoid is used to create a tile pattern as shown at the right. Describe the transformations that will map the shaded figure onto each of the positions 1–4.

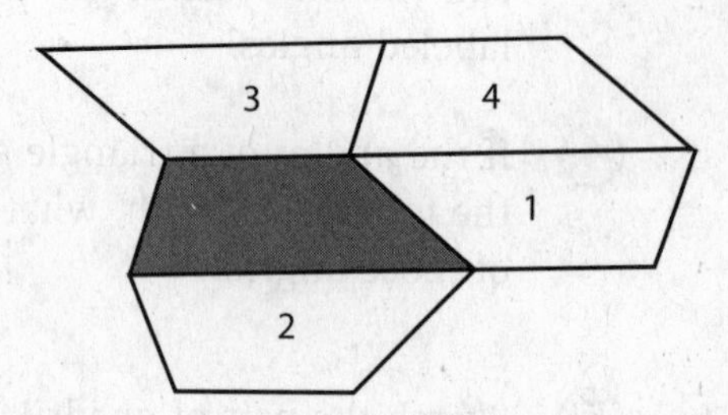

4 Describe all line and rotational symmetries found in each of the following figures.

a.

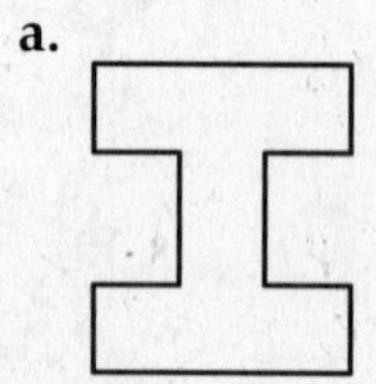

b.

Name ______________________ Date ________ Period ________

Spiral Review

Unit 6, Lesson 3

1 Kandy's Bar chocolate is packaged in triangular prism shaped boxes.

a. Draw two different possible nets that could be used to manufacture a Kandy's Bar box.

b. If the rectangular faces each measure 1.5 in. by 8 in., calculate the minimum amount of cardboard necessary to manufacture each Kandy's Bar box and the volume of each box.

2 Consider all prisms with a rectangular base of 20 inches by 15 inches.

a. Write an equation showing how the volume V of such prisms is a function of height h.

b. Find the box height that will give a volume of 4,500 cubic inches.

3 Identify the figure that each set of views represents.

a.

front view 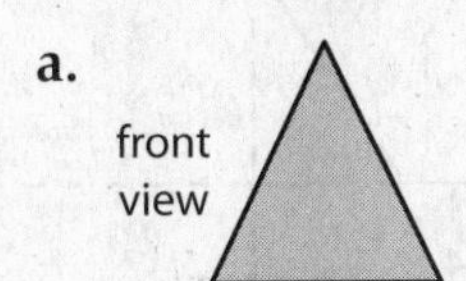 side view  top view

b.

front view side view  top view 

Review (Unit 4, Lesson 2)

4 Use what you know about geography, or consult an atlas, to make a vertex-edge graph showing which of the following western states in the United States have borders in common: Alaska (AK), Arizona (AZ), California (CA), Idaho (ID), Nevada (NV), Oregon (OR), Utah (UT), and Washington (WA). Then determine the minimum number of colors required to color a map of these states so that no adjoining states are the same color.

Name ______________________ Date ________ Period ________

Spiral Review

Unit 6, Lessons 1, 2, and 3

1 In the figure at the right, find the measure of each angle marked with a letter.

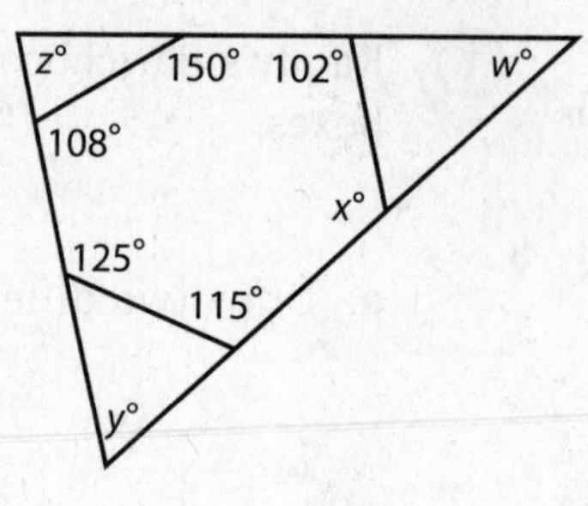

2 The sketch at the right shows the tower of a radio station with two support wires attached.

a. Find length A and length B.

b. Suppose that the radio station is able to broadcast its signal with good quality over a region with a radius of 15 miles. What is the area of that region?

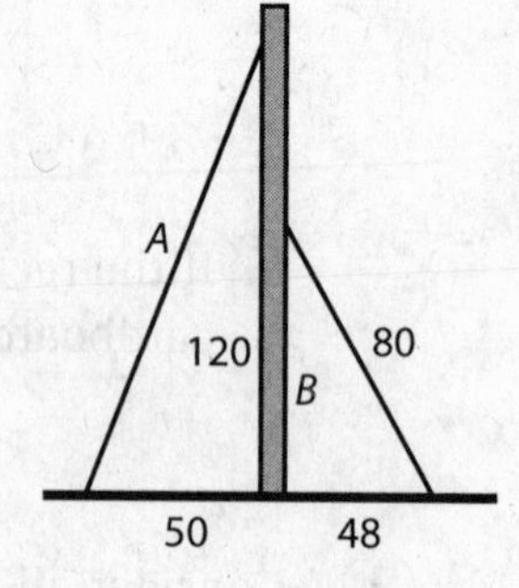

3 Examine the net of a solid shown at the right.

a. Sketch the solid and find its volume.

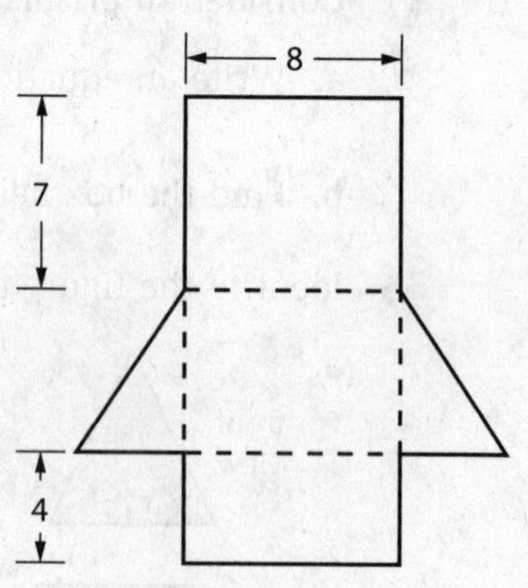

b. What is the surface area of the solid?

c. Sketch another possible net for this solid.

Review (Unit 3, Lessons 2 & 3)

4 Solve each inequality for x.

a. $5x - 7 > 33$ **b.** $3x + 4 \geq 7(x - 1)$ **c.** $5x - 6 < 2x + 3$

Name ______________________ Date ______ Period ______

Spiral Review

Unit 7, Lesson 1

1. A model rocket is launched from a height of 4 feet with an upward velocity of 64 feet per second. Its height, h, in feet after t seconds is given by the rule $h = 4 + 64t - 16t^2$.

   a. If the manufacturer wants the parachute to come out when the rocket is at its maximum height, at what time should it come out?

   b. How high will the rocket be when the parachute comes out?

2. Sketch each of the following equations without using a calculator. Use a separate piece of graph paper.

   a. $y = x^2 + 2$ **b.** $y = -x^2 - 4$ **c.** $y = 2x^2 + 1$

3. A place-kicker can kick a football off the ground with an initial upward velocity of 48 feet per second.

   a. Write a rule representing the relationship between height of the ball, h, and time, t.

   b. Make a table and a graph of values (*time, height*).

   c. At what time will the ball reach its maximum height? What height will that be?

Review (Unit 6, Lesson 1)

4. On his way to school each day, Mike can walk around the park or on a diagonal from one corner to the other, as shown on this sketch. How much distance will Mike save by walking through the park?

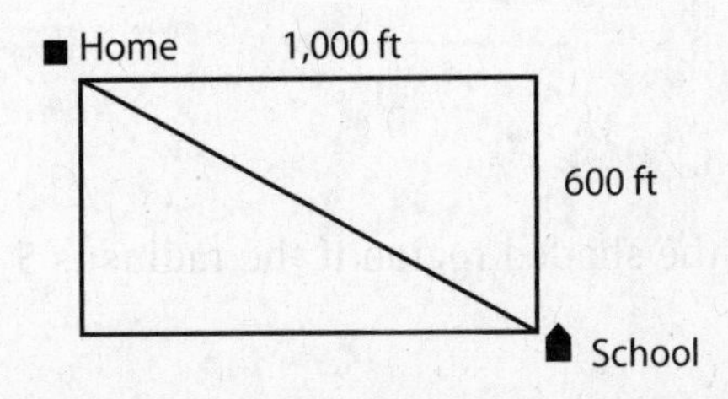

Name________________________________ Date________ Period________

Spiral Review

Unit 7, Lesson 2

1 Write each of the following quadratic expressions in equivalent standard form.

a. $a(4a + 3)$ **b.** $-c(11c + 4)$ **c.** $3m(3m + 6) - 3(m^2 + 4m + 1)$

2 Write each of the following quadratic expressions in equivalent form as the product of two linear factors.

a. $4x^2 + 8x$ **b.** $3t - 9t^2$ **c.** $-5d - 7d^2$ **d.** $cx - dx^2$

3 Write each of these quadratic expressions in equivalent forms—one expanded and one factored—so that both are as short as possible.

a. $4(2d - d^2) + 4d$

b. $-2(4s^2 - s) - (4s + 3)6s$

4 Expand each of the following products to equivalent expressions in standard quadratic form.

a. $(m + 4)(m + 1)$ **b.** $(x + 2)(x + 2)$ **c.** $(b - 3)(b + 4)$

5 Use the figure to the right to complete the following.

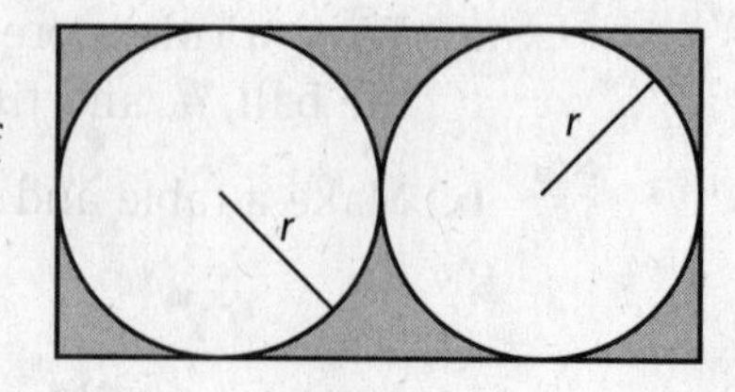

a. Write an expression for calculating the length and width of the rectangle in terms of r.

b. Write an expression for calculating the area of the rectangle.

c. Write an expression for calculating the area of one circle.

d. Write an equation for finding the area, A, of the shaded region based on the radius, r, of the circle.

e. Find the area of the shaded region if the radius is 5 cm.

Name ____________________ Date ________ Period ________

Spiral Review

Unit 7, Lesson 3

1. Solve each of the following quadratic equations by using only arithmetic operations and square roots. Show the steps of your solution process.

   a. $2m^2 + 5 = 65$ **b.** $s^2 - 4 = 12$

2. Solve each of the following equations algebraically. Show your work.

   a. $3x^2 + 15x = 0$ **b.** $-2x^2 + 5x = 0$ **c.** $4x^2 - 24x = 0$

3. Find coordinates of the maximum or minimal point on the graph of each of these quadratic functions. Tell whether the point is a maximum or a minimum.

   a. $y = 3x^2 + 15x$ **b.** $y = -2x^2 + 5x$

4. Using the quadratic function $y = -2x^2 + 8x - 5$, choose values of y to write equations that have the prescribed number of solutions. In each case, show on a graph how the condition is satisfied.

   a. Two solutions

   b. One Solution

   c. No Solutions

5. Solve each equation by using the quadratic formula. Round to the nearest tenth if necessary.

   a. $x^2 - 3x + 2 = 0$ **b.** $m^2 - 8m = -16$ **c.** $y^2 - 8y - 9 = 0$

6. The following data shows LeKishia's test scores in Ms. Rodriguez's class: 95, 47, 98, 72, 55, 88, 72, 100, 96, 72, 84, 69, 78, 84, and 90. Draw a box plot of this data. Using specific information from the box plot, explain why LeKeshia might be pleased with her performance.

Name____ Date________ Period________

Spiral Review

Unit 7, Lessons 1–3

1. Sketch a graph showing the general shape of each of the general equations below. Use a separate piece of graph paper.

   a. $y = ab^x$, $a > 0$ and $b < 1$

   b. $y = a + bx$, $a > 0$ and $b < 0$

   c. $y = ax^2$, $a > 0$

2. Write each of the following expressions in equivalent standard quadratic form.

   a. $(k + 3)(k - 8)$ **b.** $10 - b(b + 5) - (b^2 - 3b + 2)$

3. Write each of the following expressions in equivalent factored form.

   a. $3x^2 - 9x$ **b.** $w^2 + 4w$ **c.** $-2z^2 + 6z$ **d.** $12d^2 - 15d$

4. Solve each of the following equations for x using algebraic methods. Show your work. Then check your solutions using graphing technology.

   a. $2x - 3 - 3(x + 4) = 14$ **b.** $2^x = \frac{1}{4}$ **c.** $-3x^2 + 89 = 56$

5. Using algebraic methods, find the x-intercepts and the coordinates of the maximum or minimum point of the graph of each quadratic function. Then sketch a graph of each function on a separate piece of graph paper. Show your work.

   a. $y = 2x^2 - 4$ **b.** $y = x^2 + 6x$

Name ______________________ Date ______ Period ______

Spiral Review

Unit 8, Lesson 1

1. A cube has six numbers written on the six faces. The numbers are not necessarily different. Given the following information about rolling this cube, find the numbers marked on the six faces.

   Probability of rolling a 1 is $\frac{1}{2}$. Probability of rolling a 3 is $\frac{1}{3}$. It is possible to roll a 5.

2. Consider the two spinners pictured to the right.

   Spinner A

   1 2 4 3

   Spinner B

   1 2

   a. Make a chart that shows the sample space of all possible outcomes when you spin both spinners.

   b. Suppose you find the sum of the two numbers. How many possible outcomes are there? Are they equally likely?

   c. Make a probability distribution table for the sum of the two numbers.

3. Suppose you flip a penny, a nickel, and a dime and then note which come up heads and which come up tails.

   a. Make a chart that shows the sample space of all possible outcomes.

   b. What is the probability that no coin will be heads up?

Review (Unit 6, Lesson 3)

4. Draw a net that can be folded to make a rectangular box that is $3 \times 4 \times 2$ centimeters.

Name_________________________ Date________ Period________

Spiral Review

Unit 8, Lesson 1

1 Two coins are randomly chosen from a set of 1 penny, 1 nickel, 1 dime, and 1 quarter. (The first coin chosen is replaced before the second coin is chosen.)

a. Make a chart showing the sample space of all possible outcomes.

b. Make a probability distribution table for the total dollar value of the coins chosen.

c. What is the probability that the total value will be at least $0.30?

2 The following chart shows how many of the 1,000 students at Rydell High School participate in spring sports. (No students are allowed to play more than one sport in a season.)

	Baseball/ Softball	Track	Tennis
Boys	75	162	54
Girls	78	180	36

Suppose that you select a student at random from Rydell High.

a. Find the probability that the student plays tennis.

b. Find the probability that the student plays tennis or runs track.

c. Find the probability that the student is a girl who plays spring sports.

d. Find the probability that the student plays softball.

e. Find the probability that the student is a girl or plays softball. Can you find the answer using just your answers to Parts c and d? Why or Why not?

Review (Units 3, 5 and 7)

3 Solve the following equations for x.

a. $12x + 15 = 1,743$ **b.** $12x^2 + 15 = 1,743$ **c.** $12^x + 15 = 1,743$

Name ______________________________________ Date ______ Period ________

Spiral Review

Unit 8, Lesson 2

1 Suppose six students are to be chosen at random for interviews by a visiting team evaluating their school. Twenty percent of the students in the school are on the honor roll. Design and use an appropriate simulation to approximate the probability distribution for the number of honor students in the interview group.

2 Suppose that the captain of the girls varsity basketball team is a 75% free-throw shooter.

a. Each time she is sent to the line for a free throw, what is the probability that she will miss?

b. Describe how you would design one run of a simulation model for the number she makes out of 10 free throws.

c. Conduct 20 runs of your simulation and use the results to determine, on average, how many free throws out of 10 one could expect her to make.

3 Suppose that you select two numbers at random from between 0 and 5. Draw a geometric diagram and use it to find the following probabilities.

a. What is the probability that both are less than 3.5?

b. What is the probability that their sum is less than 4?

Name____ Date________ Period__________

Spiral Review

Unit 8, Lesson 2

1 William is taking a ten-item multiple-choice quiz. Each item has 5 possible choices. Since he didn't study, he chooses an answer for each question at random without reading the quiz.

a. Describe a simulation that could be used to investigate his chances of getting various scores on the quiz.

b. Fifty trials simulating this quiz were conducted. The frequency table below shows the number of questions William answered correctly in each run.

Questions Correct	0	1	2	3	4	5	6	7	8	9	10
Frequency	4	13	15	10	4	0	1	1	2	0	0

Use the frequency table to estimate the expected number of questions William will answer correctly.

c. If at least 6 questions must be answered correctly to receive a passing grade, use the results of the simulation to estimate the probability that William will receive a passing grade.

2 Jose, Claudia, and Kathleen want to explore the probability that they will all have Mrs. Parks for their Math 2 teacher next year. Students are randomly assigned to teachers, and Mrs. Parks teaches one of the four sections of Math 2.

a. Design a simulation for assigning the three students to Math 2 sections.

b. Do 25 runs of your simulation and complete a copy of the frequency table on the right.

c. Construct a histogram of your results on a separate piece of paper.

d. What is the probability that all three students will get Mrs. Parks for Math 2?

Number of Students Who Get Mrs. Parks	Frequency
0	
1	
2	
3	
Total Number of Trials	25

Review (Unit 6, Lesson 2)

3 Draw all lines of symmetry for each figure.

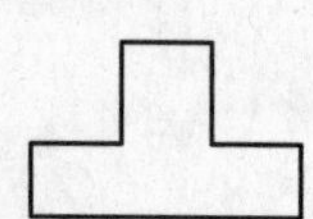

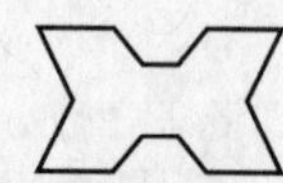

Name ______________________________ Date ________ Period ________

Standardized Test Practice

Part 1: Multiple Choice

Instructions: Fill in the appropriate circle for the best answer.

1. Which equation is always true?

A $6(x + y) = 6(y + x)$ C $6(x + y) = 6x + y$

B $6(xy) = (6x)(6y)$ D $6(x + 0) = 6x + 6$

1. Ⓐ Ⓑ Ⓒ Ⓓ

2. Solve $-17 = r + 32$.

F 49 G 15 H -15 J -49

2. Ⓕ Ⓖ Ⓗ Ⓙ

3. Solve $\frac{h}{-18} = -6$.

A -3 B 3 C 98 D 108

3. Ⓐ Ⓑ Ⓒ Ⓓ

4. The area of each square in the figure is 9 square units. What is the perimeter?

F 108 units H 36 units

G 54 units J 6 units

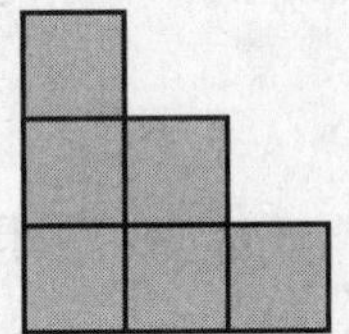

4. Ⓕ Ⓖ Ⓗ Ⓙ

5. Find the prime factorization of 240.

A $24 \cdot 10$ B $16 \cdot 15$ C $2^3 \cdot 3^2 \cdot 5$ D $2^4 \cdot 3 \cdot 5$

5. Ⓐ Ⓑ Ⓒ Ⓓ

6. Simplify $\frac{32}{48}$.

F $\frac{4}{6}$ G $\frac{2}{3}$ H $\frac{16}{24}$ J $\frac{8}{12}$

6. Ⓕ Ⓖ Ⓗ Ⓙ

7. Write $10 - 20x$ in factored form.

A $5(2 - 4x)$ B $x(10 - 20)$ C $10(1 - 2x)$ D $5(4 - 2x)$

7. Ⓐ Ⓑ Ⓒ Ⓓ

8. Which number is prime?

F 45 G 71 H 81 J 117

8. Ⓕ Ⓖ Ⓗ Ⓙ

9. Evaluate $x^2 - y^2$ if $x = 3$ and $y = -5$.

A 34 B 8 C -16 D -22

9. Ⓐ Ⓑ Ⓒ Ⓓ

10. Express $0.\overline{6}$ as a fraction in simplest form.

F $\frac{2}{3}$ G $\frac{66}{100}$ H $\frac{66}{10}$ J $\frac{11}{33}$

10. Ⓕ Ⓖ Ⓗ Ⓙ

11. State the next term in the sequence 6, 14, 22, 30, ...

A 32 B 34 C 38 D 42

11. Ⓐ Ⓑ Ⓒ Ⓓ

12. Which decimal is equivalent to $\frac{3}{8}$?

F 3.8 G 2.66 H 0.375 J 0.33

12. Ⓕ Ⓖ Ⓗ Ⓙ

13. Which is *less than* $3\frac{1}{6}$?

A 0.3166 B $3.1\overline{66}$ C $3\frac{2}{9}$ D $3\frac{1}{3}$

13. Ⓐ Ⓑ Ⓒ Ⓓ

Name ______________________ Date ________ Period ________

Standardized Test Practice *(continued)*

14. Find $1\frac{4}{7} \cdot \left(-\frac{2}{3}\right)$. Write in simplest form.

F $1\frac{1}{3}$ **G** $\frac{22}{21}$ **H** $-1\frac{1}{21}$ **J** $-2\frac{5}{14}$

14. Ⓕ Ⓖ Ⓗ Ⓙ

15. What is the equivalent measure of a pitcher containing $\frac{3}{4}$ of a gallon?

A 3 pints **B** 2 quarts **C** 10.5 cups **D** 12 cups

15. Ⓐ Ⓑ Ⓒ Ⓓ

16. What is $\frac{4}{9}$ divided by $1\frac{2}{3}$?

F $3\frac{3}{4}$ **G** $3\frac{2}{3}$ **H** $\frac{20}{27}$ **J** $\frac{4}{15}$

16. Ⓕ Ⓖ Ⓗ Ⓙ

17. Jaime works at the carwash for $3\frac{1}{2}$ hours and earns $24.50. What is his hourly wage?

A $7.50/hr **B** $7.00/hr **C** $6.50/hr **D** $6.00/hr

17. Ⓐ Ⓑ Ⓒ Ⓓ

18. Solve $\frac{x}{-5} + 4 = 24$.

F 100 **G** 4 **H** −100 **J** −140

18. Ⓕ Ⓖ Ⓗ Ⓙ

19. Find the LCM of 12 and 30.

A 6 **B** 30 **C** 60 **D** 360

19. Ⓐ Ⓑ Ⓒ Ⓓ

Part 2: Griddable

Instructions: Enter your answers by writing each digit of the answer in a column box and then shading in the appropriate circle that corresponds to that entry.

20. Evaluate $3x - 2y + 4z$ if $x = 7$, $y = 4$, and $z = 8$.

			.		
0	0	0	0	0	0
1	1	1	1	1	1
2	2	2	2	2	2
3	3	3	3	3	3
4	4	4	4	4	4
5	5	5	5	5	5
6	6	6	6	6	6
7	7	7	7	7	7
8	8	8	8	8	8
9	9	9	9	9	9

21. How much less is $\frac{1}{5}$ than $\frac{3}{4}$ in decimal form?

			.		
0	0	0	0	0	0
1	1	1	1	1	1
2	2	2	2	2	2
3	3	3	3	3	3
4	4	4	4	4	4
5	5	5	5	5	5
6	6	6	6	6	6
7	7	7	7	7	7
8	8	8	8	8	8
9	9	9	9	9	9

Name _______________ Date _______ Period _______

Standardized Test Practice *(continued)*

Part 3: Short Response

Instructions: Write your answer in the blank at the right of each question.

22. BOOK FAIR Eva and Laura spent a total of $33 at a book fair. Eva spent $5 more than Laura. Write and solve an equation to find how much Laura spent at the book fair. **22.** _______________

23. Find the mean, median, and mode for the set of data shown in the line plot. If necessary, round to the nearest tenth. **23.** _______________

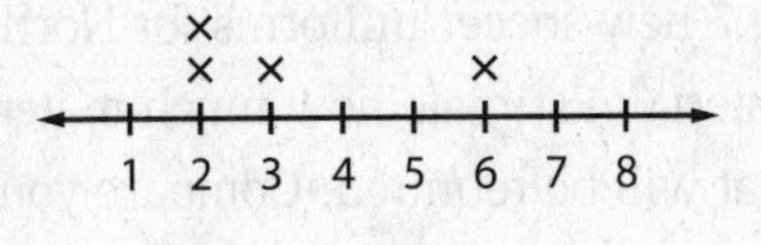

Write the fraction in simplest form.

24. $\frac{39}{72}$ **24.** _______________

25. $\frac{16a^2b}{24ab}$ **25.** _______________

Find the product or quotient.

26. $m^3 \cdot m \cdot m^5$ **26.** _______________

27. $(4x^4)(7xy^3)$ **27.** _______________

28. $\frac{-4b^7}{2b^5}$ **28.** _______________

29. Find the mean of the test scores 60, 80, 85, 85, and 95. **29.** _______________

30. Write 3.65 as a fraction in simplest form. **30.** _______________

31. Write the next three terms of the sequence 4, 6.5, 9, 11.5, ... **31.** _______________

32. Write $\frac{3}{11}$ as a decimal. Use a bar to show a repeating decimal. **32.** _______________

33. Solve $b = \frac{5}{7} + \frac{2}{21}$. **33.** _______________

34. Laurie is making muffins. Her muffin recipe calls for $3\frac{1}{2}$ cups of flour.

a. If she has $10\frac{1}{2}$ cups of flour, how many batches of the muffin recipe can she make? **34a.** _______________

b. If she plans on making 5 batches of the muffin recipe, how many cups of flour does she need? **34b.** _______________

Name ________________________ Date ________ Period ________

Standardized Test Practice *(continued)*

Score ________

Part 4: Extended Response

Demonstrate your knowledge by giving a clear, concise solution to each problem. Be sure to include all relevant drawings and justify your answers. You may show your solution in more than one way or investigate beyond the requirements of the problem.

35. Seams and Hems is a custom sewing center. The company does sewing for individuals and organizations.

a. The sewing center is making 7 new soccer uniforms for Northmont High School. Each uniform will require about $4\frac{5}{8}$ yards of material. Estimate how much material will be required. Then compute the actual amount of material that will be required. Compare your answer with your estimate. Explain how you know your answer is reasonable.

b. Mr. Ortiz has 40 yards of material that he wants made into aprons. Each apron uses $1\frac{3}{4}$ yards of material. Find how many aprons can be made. Explain whether the answer needs to be rounded up or down and why. Show any diagrams or computations that may help your explanation.

36. Bob makes quilts in his free time.

a. Bob is making 3 baby quilts that will require $4\frac{1}{2}$ yards of fabric each. Determine how much material he will need.

b. Bob is making quilted wall hangings that will require $2\frac{1}{3}$ yards of fabric each. He has 14 yards of fabric. How many wall hangings can he make?

Name ____________________ Date ________ Period ________

Standardized Test Practice

Score ________

Part 1: Multiple Choice

Instructions: Fill in the appropriate circle for the best answer.

1. Evaluate $x + y - z$ if $x = 3$, $y = 2$, and $z = 4$.

A -3 B 1 C 5 D 9

1. Ⓐ Ⓑ Ⓒ Ⓓ

2. Solve $\frac{x}{-7} = -196$.

F 1372 G 28 H -28 J -189

2. Ⓕ Ⓖ Ⓗ Ⓙ

3. Write $8b + 20$ in factored form.

A $4(b + 5)$ B $8(b + 5)$ C $4(2b + 5)$ D $2(4b + 10)$

3. Ⓐ Ⓑ Ⓒ Ⓓ

4. Find the least common multiple (LCM) of $16st^2u$ and $8s^2u^2$.

F $8stu$ G $16s^2t^2u^2$ H $128s^3t^2u^3$ J $2st^2u^2$

4. Ⓕ Ⓖ Ⓗ Ⓙ

5. Find the sum of $3\frac{4}{5}$ and $2\frac{7}{10}$. Write in simplest form.

A $6\frac{5}{10}$ B $6\frac{1}{2}$ C $5\frac{11}{15}$ D $5\frac{1}{2}$

5. Ⓐ Ⓑ Ⓒ Ⓓ

6. How much less is $\frac{7}{16}$ than $2\frac{1}{4}$?

F $1\frac{13}{16}$ G $1\frac{3}{4}$ H $1\frac{1}{2}$ J $\frac{15}{16}$

6. Ⓕ Ⓖ Ⓗ Ⓙ

7. Which scale has a scale factor of $\frac{1}{48}$?

A 2 in. = 36 ft B 3 in. = 12 ft C 2 in. = 6 ft D 3 in. = 6 ft

7. Ⓐ Ⓑ Ⓒ Ⓓ

8. Amelia paid \$4.79 for 2 gallons of juice. What was the price per quart of juice?

F \$2.40 G \$1.20 H \$0.60 J \$0.40

8. Ⓕ Ⓖ Ⓗ Ⓙ

9. Choose the best estimate for 48% of 438.

A 175 B 220 C 240 D 260

9. Ⓐ Ⓑ Ⓒ Ⓓ

10. The graph shows the results of a survey on voting preferences. Out of a group of 500 voters, how many would you expect to say they prefer to vote on the Internet?

F 24 H 120

G 115 J 200

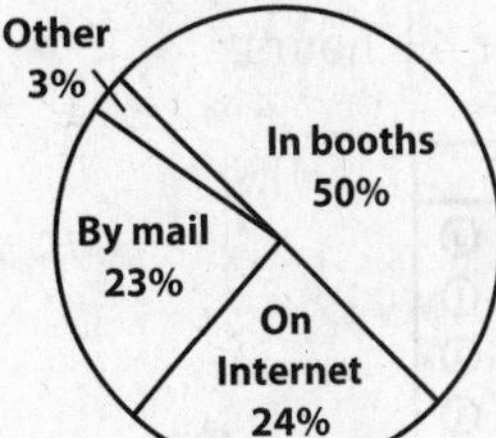

10. Ⓕ Ⓖ Ⓗ Ⓙ

Name ______________________ Date ________ Period ________

Standardized Test Practice *(continued)*

11. Solve $4m + 6 = 8m - 2$.

A -2 B -1 C 1 D 2 **11.** Ⓐ Ⓑ Ⓒ Ⓓ

12. Write the inequality for the graph.

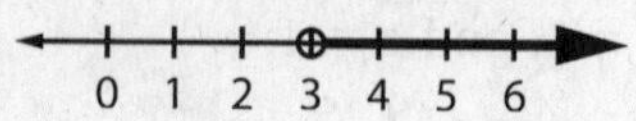

F $b \geq 3$ G $b > 3$ H $b \leq 3$ J $b < 3$ **12.** Ⓕ Ⓖ Ⓗ Ⓙ

13. Solve $x - 6.9 \geq -9.1$.

A $x \geq 2.2$ B $x \geq -2.2$ C $x \geq -16$ D $x \leq -2.2$ **13.** Ⓐ Ⓑ Ⓒ Ⓓ

14. Solve $\frac{m}{-6} \leq -30$.

F $m < 30$ G $m \leq 180$ H $m > 30$ J $m \geq 180$ **14.** Ⓕ Ⓖ Ⓗ Ⓙ

15. Which inequality represents *six less than three times a number is more than eighteen?*

A $(6 - 3)n > 18$ C $3n - 6 > 18$

B $3n + 6 > 18$ D $6n - 3 > 18$ **15.** Ⓐ Ⓑ Ⓒ Ⓓ

Simplify each expression.

16. $-6(-4a)(-2b)$.

F $48ab$ G $-12ab$ H $-48ab$ J -48 **16.** Ⓕ Ⓖ Ⓗ Ⓙ

17. $4t + 4 - 11 + t$

A $5t - 7$ B $3t - 7$ C $-2t$ D $5t + 15$ **17.** Ⓐ Ⓑ Ⓒ Ⓓ

Part 2: Griddable

Instructions: Enter your answer by writing each digit of the answer in a column box and then shading in the appropriate circle that corresponds to that entry.

18. Find the number of miles traveled by driving at 50 miles per hour for $4\frac{1}{2}$ hours.

				.		
0	0	0	0		0	0
1	1	1	1		1	1
2	2	2	2		2	2
3	3	3	3		3	3
4	4	4	4		4	4
5	5	5	5		5	5
6	6	6	6		6	6
7	7	7	7		7	7
8	8	8	8		8	8
9	9	9	9		9	9

19. The perimeter of a rectangle is 52 feet. The length is 2 feet shorter than three times the width. Find the area of the rectangle in square feet.

				.		
0	0	0	0		0	0
1	1	1	1		1	1
2	2	2	2		2	2
3	3	3	3		3	3
4	4	4	4		4	4
5	5	5	5		5	5
6	6	6	6		6	6
7	7	7	7		7	7
8	8	8	8		8	8
9	9	9	9		9	9

Name ________________________ Date ______ Period ______

Standardized Test Practice *(continued)*

Part 3: Short Response

Instructions: Write your answer in the blank at the right of each question.

20. A new mountain bike is worth only about 0.70 of its value from the previous year during the first four years after it is purchased. Approximately how much will a $900 bicycle be worth in 1, 2, 3, and 4 years? Round your answers to the nearest whole dollar.

20. ______________

21. The number of baskets scored by a high school basketball team are shown in the table. Write an inequality that represents the number of baskets they must score in their sixth game to have an average of more than 40 baskets per game.

Game	Baskets
1	29
2	43
3	36
4	46
5	38

21. ______________

22. Find $3\frac{1}{3} \div 2\frac{3}{4}$.

22. ______________

23. Find the least common denominator (LCD) of $\frac{1}{16a^2b}$ and $\frac{5}{24ab}$.

23. ______________

24. Find $3\frac{5}{6} - 5\frac{2}{3}$.

24. ______________

25. What is the scale factor of the scale $\frac{1}{2}$ inch $= 2$ feet?

25. ______________

26. Express $\frac{7}{16}$ as a percent.

26. ______________

27. Solve $5(a + 3) - 4 = 3a + 27$.

27. ______________

28. Belinda has more than two times as many shoes as Georgia and more than twice as many as Nina. If both Georgia and Nina have at least 5 pairs each, how many pairs of shoes does Belinda have?

28. ______________

29. Five times the sum of a number and 15 is greater than 37. What is the number?

29. ______________

30. Solve $x + \frac{3}{5} > 4$.

30. ______________

31. Suppose a place setting of china is on sale at a 15% discount. It normally sells for $129.99.

a. Write an equation that can be used to determine the sale price of the china.

31a. ______________

b. Find the sale price.

31b. ______________

c. What will the sale price be if 5% tax is added to the cost?

31c. ______________

Name ______________________ Date ________ Period ________

Standardized Test Practice *(continued)*

Score ________

Part 4: Extended Response

Demonstrate your knowledge by giving a clear, concise solution to each problem. Be sure to include all relevant drawings and justify your answers. You may show your solution in more than one way or investigate beyond the requirements of the problems.

32. Make up a problem that can be solved by using the inequality $\$18.00 - 2p \geq \$6.00$. Then solve the inequality and graph it on a number line. What does the solution represent?

33. Make up a problem that can be solved by using the equation $\$2.43 = \$15 - 3x$. Then solve the equation. What does the answer represent?

34. At an office supply store, a box of 60 pens contains pens of four different colors. It contains five times as many red pens as black pens, and four more black pens than green pens. The number of red, green, and black pens combined is three times the number of blue pens. How many pens of each color are in the box?

35. Mr. Rodriguez needs to keep the weight of a package he is sending to his sister and three children under 4 pounds.

He wishes to send three identical calculators and a $1\frac{3}{4}$-pound chess set.

a. Write an inequality that describes the situation.

b. What does the variable represent?

c. Solve the inequality.

d. What does the solution to the inequality represent?

Name ________________ Date ______ Period ______

Standardized Test Practice

Score ______

Part 1: Multiple Choice

Instructions: Fill in the appropriate circle for the best answer.

1. Which verbal expression represents the phrase *nine less than seven times a number*?

A $7n - 9$ **B** $n - 9$ **C** $9 - 7n$ **D** $7 + n - 9$

1. Ⓐ Ⓑ Ⓒ Ⓓ

2. Write $0.\overline{36}$ as a fraction.

F $\frac{4}{9}$ **G** $\frac{4}{11}$ **H** $\frac{36}{100}$ **J** $\frac{1}{36}$

2. Ⓕ Ⓖ Ⓗ Ⓙ

3. Solve $-9x \leq -54$.

A $x \geq 6$ **B** $x \leq 6$ **C** $x \leq -4$ **D** $x \geq -6$

3. Ⓐ Ⓑ Ⓒ Ⓓ

4. Find the slope of the line that passes through the points $A(-3, -5)$ and $B(1, -1)$.

F $\frac{3}{2}$ **G** 1 **H** $-\frac{2}{3}$ **J** -1

4. Ⓕ Ⓖ Ⓗ Ⓙ

5. What is the scale in a scale drawing where 3 inches is 15 feet?

A $\frac{1 \text{ in.}}{12 \text{ ft}}$ **B** $\frac{1 \text{ in.}}{5 \text{ ft}}$ **C** $\frac{1 \text{ in.}}{4 \text{ ft}}$ **D** $\frac{5 \text{ in.}}{1 \text{ ft}}$

5. Ⓐ Ⓑ Ⓒ Ⓓ

6. Which set of numbers represents the lengths of the sides of a right triangle?

F 4, 4, 9 **G** 5, 9, 12 **H** 12, 16, 20 **J** 6, 7, 13

6. Ⓕ Ⓖ Ⓗ Ⓙ

7. Find the distance between the points $Q(10, -8)$ and $R(-15, 7)$. Round to the nearest tenth.

A 14.1 **B** 15.8 **C** 20.0 **D** 29.2

7. Ⓐ Ⓑ Ⓒ Ⓓ

8. Frank draws two similar triangles. One triangle has sides of 8, 9, and 10 centimeters. The shortest side of the second triangle is 20 centimeters long. Find the length of the longest side of the second triangle.

F 28 cm **G** 26 cm **H** 25 cm **J** 22 cm

8. Ⓕ Ⓖ Ⓗ Ⓙ

9. Identify the transformation in the graph at the right.

A translation **C** rotation

B reflection **D** horizontal

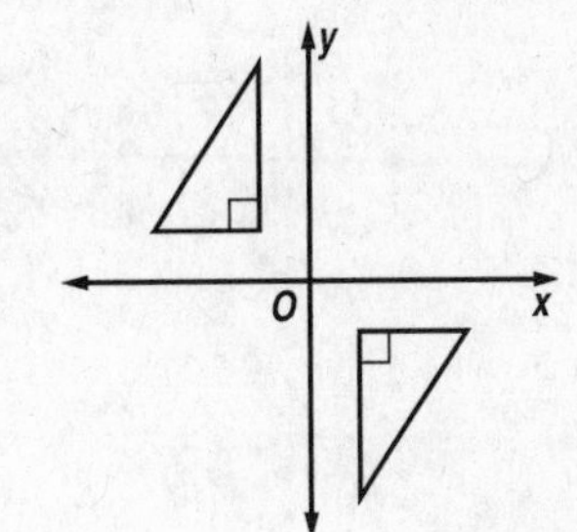

9. Ⓐ Ⓑ Ⓒ Ⓓ

Standardized Test Practice *(continued)*

10. Find the area of a trapezoid with bases of 12 inches and 10 inches and a height of 15 inches.

F 330 in^2 **G** 165 in^2 **H** 74 in^2 **J** 66 in^2 **10.** Ⓕ Ⓖ Ⓗ Ⓙ

11. Find the radius of a circle if its circumference is 50.24 meters.

A 4 m **B** 8 m **C** 16 m **D** 32 m **11.** Ⓐ Ⓑ Ⓒ Ⓓ

12. Identify the solid.

F triangular prism

G square pyramid

H cone

J triangular pyramid

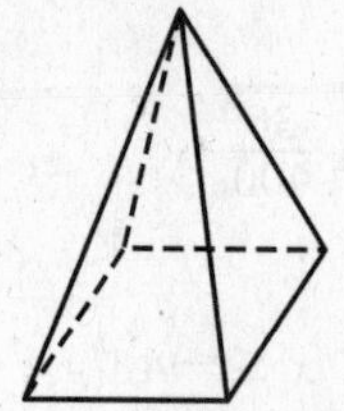

12. Ⓕ Ⓖ Ⓗ Ⓙ

13. Find the volume of a rectangular prism with length of 12 inches, width of 9 inches, and height of 6 inches.

A 648 in^3 **B** 468 in^3 **C** 324 in^3 **D** 126 in^3 **13.** Ⓐ Ⓑ Ⓒ Ⓓ

14. Roger has a cone-shaped cotton candy container that is 12 inches high and has a radius of 6 inches. Find the volume of the container.

F 72 in^3 **G** 432 in^3 **H** 452.4 in^3 **J** 1357.2 in^3 **14.** Ⓕ Ⓖ Ⓗ Ⓙ

15. Find the product of $(-11r^2s^7)(3rst^4)$.

A $-33rst$ **B** $33r^3s^8t^4$ **C** $-33r^3s^8t^4$ **D** $-33r^2s^7t^4$ **15.** Ⓐ Ⓑ Ⓒ Ⓓ

16. Solve $9x - 7 > 47$.

F $x > 360$ **G** $x = 6$ **H** $x < 6$ **J** $x > 6$ **16.** Ⓕ Ⓖ Ⓗ Ⓙ

Part 2: Griddable

Instructions: Enter your answer by writing each digit of the answer in a column box and then shading in the appropriate circle that corresponds to that entry.

17. Find the 6th term of the sequence 19, 16, 13, 10, … .

				.		
0	0	0	0		0	0
1	1	1	1		1	1
2	2	2	2		2	2
3	3	3	3		3	3
4	4	4	4		4	4
5	5	5	5		5	5
6	6	6	6		6	6
7	7	7	7		7	7
8	8	8	8		8	8
9	9	9	9		9	9

18. Find the surface area in square inches of a soup can that has a diameter of 3 inches and a height of $4\frac{1}{2}$ inches. Round to the nearest tenth.

				.		
0	0	0	0		0	0
1	1	1	1		1	1
2	2	2	2		2	2
3	3	3	3		3	3
4	4	4	4		4	4
5	5	5	5		5	5
6	6	6	6		6	6
7	7	7	7		7	7
8	8	8	8		8	8
9	9	9	9		9	9

Name ______________________ Date ________ Period ________

Standardized Test Practice *(continued)*

Part 3: Short Response

Instructions: Write your answer in the blank at the right of each question.

19. Find the percent of change, to the nearest tenth, in the price of a computer from $1145 to $1420. State whether the change is an increase or decrease. **19.** ________

20. Find the distance between the points $A(-2, -3)$ and $B(5, 7)$. Round to the nearest tenth. **20.** ________

21. Replace the ● with <, > or = to make a true statement.
$-\sqrt{59}$ ● -7.9 **21.** ________

22. The lengths of the sides of a triangle are 14, 48, and 50. Is this triangle a right triangle? **22.** ________

23. What is the area of a triangle with a height of 6 inches and a base of 5 inches? **23.** ________

24. What is the area of a trapezoid with bases of 9 meters and 18 meters, and a height of 10 meters? **24.** ________

25. Find the circumference of a circle with a radius of 8 meters. Round to the nearest tenth. **25.** ________

26. What is the volume of a rectangular prism with a length of 8 meters, a width of 10 meters and a height of 20 meters? **26.** ________

27. Find the surface area of a cone with radius of 5 cm and slant height of 9 cm. Round to the nearest tenth. **27.** ________

28. Nikki has a model of an Egyptian pyramid that has a slant height of $5\frac{1}{4}$ inches and a 3-inch square base.

a. What is the surface area of the pyramid? **28a.** ________

b. What is the volume of the pyramid? Round to the nearest tenth. **28b.** ________

c. If Nikki doubles the dimensions, what are the new surface area and volume? Round to the nearest tenth if necessary. **28c.** ________

Name ______________________ Date ______ Period ______

Standardized Test Practice *(continued)*

Score ______

Part 4: Extended Response

Demonstrate your knowledge by giving a clear, concise solution to each problem. Be sure to include all relevant drawings and justify your answers. You may show your solution in more than one way or investigate beyond the requirements of the problem.

The Food and Drug Administration, among other duties, is charged with protecting consumers from misleading packaging. Many schemes have been used through the years to make customers believe they are getting more than they really are.

29. Find the volume of the cylinder at the right. Find the volume if the height of the cylinder is doubled. Find the volume if the radius is doubled. Round your answers to the correct number of significant digits. Do you think retailers would tend to use tall bottles or large-diameter bottles to mislead customers? Why or why not?

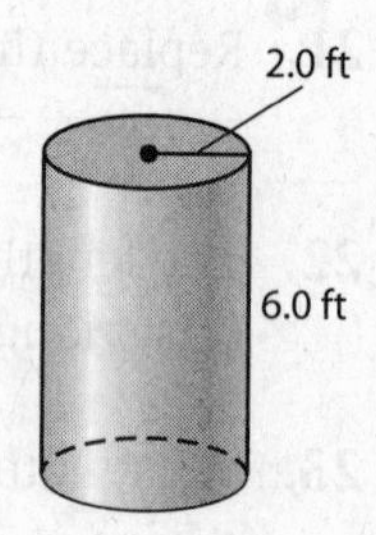

30. A way of misleading customers is by creating false bottoms such as those shown at the right. Guess which false bottom displaces more volume. Estimate the volume of each false bottom to check your answer. Give examples of other ways in which the actual volume of a container may be less than it appears to be.

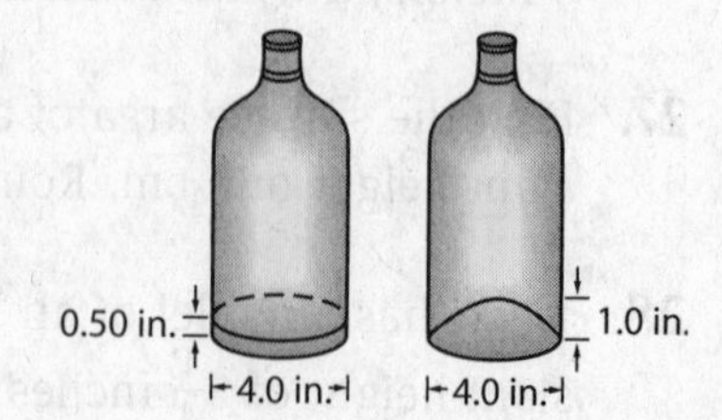

Name ______________________ Date ______ Period ______

Standardized Test Practice

Part 1: Multiple Choice

Instructions: Fill in the appropriate circle for the best answer.

1. Express 1500 in scientific notation.

A 1.5×10^2 **B** 0.15×10^3 **C** 1.5×10^3 **D** 1.5×10^{-3}

1. Ⓐ Ⓑ Ⓒ Ⓓ

2. Find the amount of simple interest earned on $500 at an annual rate of $5\frac{1}{2}$% for 3 years.

F $27.50 **G** $78 **H** $82.50 **J** $90

2. Ⓕ Ⓖ Ⓗ Ⓙ

3. Find the y-intercept of the graph of $2x - y = 10$.

A 10 **B** 2 **C** −2 **D** −10

3. Ⓐ Ⓑ Ⓒ Ⓓ

4. What are the coordinates of the midpoint of the line segment with endpoints $J(0, -6)$ and $K(8, -4)$.

F (4, −5) **G** (−2, 1) **H** (4, −1) **J** (−2, −5)

4. Ⓕ Ⓖ Ⓗ Ⓙ

5. If the measure of the hypotenuse of a right triangle is 15 meters and the measure of one leg is 9 meters, what is the measure of the other leg?

A 306 m **B** 144 m **C** 12 m **D** $\sqrt{306}$ m

5. Ⓐ Ⓑ Ⓒ Ⓓ

6. Classify the quadrilateral at the right with the name that best describes it.

F square **H** trapezoid

G quadrilateral **J** parallelogram

6. Ⓕ Ⓖ Ⓗ Ⓙ

7. Find the volume of a rectangular prism with a length of 5 meters, a width of 6 meters, and a height of 10 meters.

A 600 m^3 **B** 300 m^3 **C** 150 m^3 **D** 75 m^3

7. Ⓐ Ⓑ Ⓒ Ⓓ

8. Find the surface area of a cone with a diameter of 11 meters and a slant height of 8.6 meters.

F 148.6 m^2 **G** 243.6 m^2 **H** 392.2 m^2 **J** 677.3 m^2

8. Ⓕ Ⓖ Ⓗ Ⓙ

9. If the dimensions of a triangular prism are doubled, the volume

A stays the same. **C** is quadrupled.

B is doubled. **D** is 8 times greater.

9. Ⓐ Ⓑ Ⓒ Ⓓ

Name ______________________________ Date ________ Period ________

Standardized Test Practice *(continued)*

For Questions 10 and 11, use the box-and-whisker plot shown.

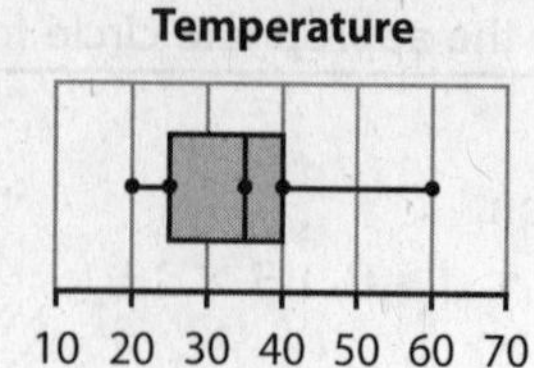

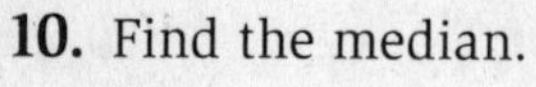

10. Find the median.

A 20 C 35
B 25 D 40

10. Ⓐ Ⓑ Ⓒ Ⓓ

11. What is the lower quartile?

F 120 G 40 H 35 J 25

11. Ⓕ Ⓖ Ⓗ Ⓙ

12. A bag contains 3 red marbles, 4 green marbles, and 2 blue marbles. Kaya chooses a marble at random, then without replacing it chooses a second marble. What is the probability that Kaya chooses two green marbles?

A $\frac{59}{72}$ B $\frac{16}{81}$ C $\frac{1}{6}$ D $\frac{4}{27}$

12. Ⓐ Ⓑ Ⓒ Ⓓ

13. The coordinates of the endpoints of a segment are $E(-6, -2)$ and $F(-4, 8)$. What are the coordinates of the midpoint of this segment?

F $(-5, 3)$ G $(-10, 6)$ H $(-1, -5)$ J $(3, -5)$

13. Ⓕ Ⓖ Ⓗ Ⓙ

14. Find the surface area of a rectangular prism with length 5 cm, width 9 cm, and height 1 cm.

A 15 cm^2 B 30 cm^2 C 45 cm^2 D 118 cm^2

14. Ⓐ Ⓑ Ⓒ Ⓓ

15. Find the interquartile range for the quiz scores 45, 25, 60, 35, 20, 40, and 55.

F 10 G 15 H 30 J 40

15. Ⓕ Ⓖ Ⓗ Ⓙ

Part 2: Griddable

Instructions: Enter your answers by writing each digit of the answer in a column box and then shading in the appropriate circle that corresponds to that entry.

16. Find the area in square feet of a triangle with a base of 24 feet and a height of 18 feet.

17. Find the value of 7!.

Name ______________________ Date ________ Period ________

Standardized Test Practice *(continued)*

Part 3: Short Response

Instructions: Write your answer in the blank at the right of each question.

18. Display the number of home runs in a stem-and-leaf plot.

18.

Top 10 Home Run Hitters NY Yankees, 2001	
Player	**Home Runs**
T. Martinez	34
B. Williams	26
J. Posada	22
D. Jeter	21
P. O'Neill	21
D. Justice	18
A. Soriano	18
B. Brosius	13
S. Spencer	10
C. Knoblauch	9

19. An algebra class had 36 students. Four of the students transferred to pre-algebra. What percent of the algebra class transferred?

19. ________________

20. Suppose y varies directly with respect to x and the constant of variation is -5. What is the rate of change of this direct variation equation?

20. ________________

21. At the same time a light pole casts a 5-foot shadow, a nearby 4.5-foot girl casts a 2-foot shadow. How tall is the light pole?

21. ________________

22. A figure has vertices $A(3, 1)$, $B(-2, 0)$, $C(0, -4)$, and D$(2, -3)$. After a translation of 3 units right and 2 units down what are the coordinates of the new vertices?

22. ________________

23. Identify a pair of skew lines in the figure at the right.

E F H G A B D C

23. ________________

24. Two six-sided number cubes are rolled.

a. Find the probability of rolling a sum of 2.

24a. ________________

b. Find the probability of a sum of 8.

24b. ________________

Name ______________________ Date ______ Period ________

Standardized Test Practice *(continued)*

Part 4: Extended Response

Demonstrate your knowledge by giving a clear, concise solution to each problem. Be sure to include all relevant drawings and justify your answers. You may show your solution in more than one way or investigate beyond the requirements of the problem.

25. Probability is often used in baseball. Suppose Juan and Tony are the first two batters in the ninth inning. Their respective batting averages are 0.200 or 20% and 0.250 or 25%.

a. Design a simulation to find the probability of both getting a hit.

b. Tell how to find the probability of independent events. Find the probability of both Juan and Tony getting a hit, assuming independence.